T0286617

Functional Analysis Revisited

Functional Analysis Revisited is not a first course in functional analysis – although it covers the basic notions of functional analysis, it assumes the reader is somewhat acquainted with them. It is by no means a second course either: there are too many deep subjects that are not within scope here. Instead, having the basics under his belt, the author takes the time to carefully think through their fundamental consequences.

In particular, the focus is on the notion of completeness and its implications, yet without venturing too far from areas where the description 'elementary' is still valid. The author also looks at some applications, perhaps just outside the core of functional analysis, that are not completely trivial. The aim is to show how functional analysis influences and is influenced by other branches of contemporary mathematics. This is what we mean by Functional Analysis Revisited.

ADAM BOBROWSKI is a professor in the Department of Mathematics at Lublin University of Technology, Poland. He was awarded the Hugo Steinhaus Prize for his achievements in analyzing mathematical models of biological reality and has authored more than 70 scientific papers and 6 books. His works include *Functional Analysis for Probability and Stochastic Processes* (2005), *Convergence of One-Parameter Operator Semigroups* (2016) and *Generators of Markov Chains* (2020).

Functional Analysis Revisited

An Essay on Completeness

ADAM BOBROWSKI
Lublin University of Technology, Poland

Shaftesbury Road, Cambridge CB2 8EA, United Kingdom

One Liberty Plaza, 20th Floor, New York, NY 10006, USA

477 Williamstown Road, Port Melbourne, VIC 3207, Australia

314–321, 3rd Floor, Plot 3, Splendor Forum, Jasola District Centre, New Delhi – 110025, India

103 Penang Road, #05–06/07, Visioncrest Commercial, Singapore 238467

Cambridge University Press is part of Cambridge University Press & Assessment, a department of the University of Cambridge.

We share the University's mission to contribute to society through the pursuit of education, learning and research at the highest international levels of excellence.

www.cambridge.org
Information on this title: www.cambridge.org/9781009430913
DOI: 10.1017/9781009430883

When citing this work, please include a reference to the DOI 10.1017/9781009430883

First published 2024

A catalogue record for this publication is available from the British Library

A Cataloging-in-Publication data record for this book is available from the Library of Congress

ISBN 978-1-009-43091-3 Hardback
ISBN 978-1-009-43089-0 Paperback

Contents

Introduction

This book has been devised for students who, having perhaps taken a first course in functional analysis but not being ready for a second, more advanced one, would like to take another look at the subject. To accommodate their needs, we do go through basic notions, but the stress on a systematic study is considerably lighter than in other monographs. Instead, our main goal is to show how the notion of *completeness* permeates functional analysis and modern mathematical analysis as a whole.

Hence, this is not a basic course in functional analysis. Neither is it a second course; we do not even dare to touch more advanced topics; we want to spend more time thinking about the basics. It is thus an in-between course, an intermediate course: 'Functional Analysis Revisited.' Perhaps a preparation for a second course, perhaps not.

At the same time, the book is a testimony to the author's fascination with the simplicity and beauty of the notion of completeness. Nearly every chapter contains a main result, an important one for a branch of mathematics, or science in general, that crucially hinges on the completeness of a metric space involved. We learn thus that, were it not for completeness, Achilles would not catch the tortoise, a functional series would not converge, a contraction mapping would not have a fixed point, and differential equation would not have a solution; neither would a renewal equation, and as a result a model of population growth would be rendered useless. To continue the list: a continuous function would not be Riemann integrable, and thus the fundamental theorem of calculus would cease to be true; polynomials would not approximate continuous functions well; we would not be able to project on subspaces of Hilbert spaces; and we would have an issue with solving the all-important heat equation. Without the notion of completeness, mathematics, as we know it today, would not exist, or at least it would be in a very pitiful state.

It is thus no surprise that Banach spaces, that is, normed linear spaces that are complete, play a central role in this little book. Banach spaces are appealing blends of algebraic and topological notions, with a structure simple enough to be found everywhere around us, and at the same time rich enough to lead to satisfactory, deep theory including the uniform boundedness principle of Banach and Steinhaus, and the open mapping and closed graph theorems. For a mathematician, they are a pleasure to look at and a joy to discover in the heart of applied problems.

***

Technically – though corrected, expanded, rearranged and rethought – the book is a translation of my *Analiza funkcjonalna jeden i pół. Szkic o zupełności*, published (in Polish) by Lublin University of Technology Press in 2015. The original has served as a textbook for several courses in functional analysis at this university, and was used by my esteemed colleagues at Silesian University and Łódź University of Technology. I am very grateful for all the encouragement and critical remarks that have made the present edition better than the original.

As compared with the Polish edition, besides new figures and pictures, new material added or corrected in certain sections (like the Hausdorff Moment Problem) and a few new sections (like the ones devoted to convergence of Fourier series, the Fejér theorem (Section 14.4) and uniform convergence (Section 15.4) in particular), there are two completely new chapters. These are Chapters 15 and 16.

In Chapter 15 we stop skimming the surface and finally go a bit deeper into the structure of complete spaces. Namely, we discuss the Baire Category theorem and see the Banach–Steinhaus uniform boundedness principle as a consequence of this fundamental result. This leads us naturally to the open mapping and closed graph theorems, and the perspective that these provide on linear operators is so appealing that we are not able to resist the temptation of subsequently touring the land of semigroups of operators in Chapter 16. I hope readers will enjoy this new material, and will find it useful in their own pursuit of mathematics.

***

Every chapter ends with a short, non-technical summary of its most important results, preceded by plenty of reasonably-pitched exercises; a few that are a little more demanding are marked with a ⛰ sign. Some of the exercises are taken directly from the original Polish edition; some were created for the sake of midterm and final exams that took place after the Polish book's publication, and thus in part are due to Adam Gregosiewicz.

I am grateful to many colleagues, including R. Bogucki and B. Przeradzki, who notified me about a number of typos in the Polish version of the book; these errors were of course corrected in this edition. I owe special thanks to W. Chojnacki for his constant efforts to make my English, and the English of this book in particular, correct and more readable (to make it simple is *mission impossible*). E. Ratajczyk diligently read the entire apparently ready-to-be-published text, and still found a great number of misprints. Due to her efforts, there is hope that the set of errors is nowhere dense in this book – at most a set of the first category.

Finally, I would like to thank the CUP team, including R. Astely, A. Jacobsen and C. Dennison for their professional support.

1
Complete Metric Spaces

MONSIEUR JOURDAIN: Well, what do you know about that! These forty years now I've been speaking in prose without knowing it!

–Molière, The Bourgeois Gentleman, 1670

1.1 Maps

Imagine that a map of the county, city or village you are living in is placed on the ground somewhere within the country's, city's or village's borders. It can be proved, and our intuition confirms this conjecture, that there is a point (precisely one!) on this map that lies directly above the point it describes. This statement is true regardless of the map scale. And of course it does not matter which country, city or village we have in mind. What is important is only that the map describes the entire area in which it is placed.

Moreover, there is nothing special about two dimensions in this example. If a one-dimensional 'map' of a road from town A to town B is prepared and placed somewhere on this road, then a point on this map can be found that lies directly above the place on the road it describes. The same is true in three dimensions: if a three-dimensional map of a lecture hall is placed in that lecture hall, there is a point in this map that is placed precisely at the point it describes.

We have thus found a common denominator for a number of 'spaces': these spaces are distinguished by the fact that their maps, when placed in the corresponding spaces, have one point lying precisely in the place it describes.

There are also, of course, spaces that do not posses this property. Think, for example, of a punched ball B, that is, of a ball with removed center, call this center O. Any smaller punched ball B', with the same center, that is contained in B, can be thought of as a map of B. Namely, if $k > 1$ is the ratio of radii of B and B', a point $P' \in B'$ can be thought of as an image of $P \in B$ if and only if $\overrightarrow{OP} = k\,\overrightarrow{OP'}$. Certainly on the map B' there is no point P' that is an image of itself. The reason for this situation is that O, the removed point, is the only candidate for having this property.

There are thus two types of spaces: those with holes and those without holes. The latter are professionally termed *complete* and the holey spaces are said to be incomplete (see further down for a more precise definition).

1.2 Roots

We have encountered complete metric spaces in mathematics a number of times before without perhaps knowing it (like Monsieur Jourdain from our Molière quote). To present an example of such an encounter, we start from the Bernoulli inequality

$$(x+1)^n \geq 1 + nx, \text{ where } x \geq -1, n \geq 1, \tag{1.1}$$

which is easy to prove by induction. We will show first, following Lech Maligranda (see [28] and the papers cited there; an almost identical proof was given even earlier by Bengt Åkerberg [3]), that (1.1) implies the following relation:[1]

$$x_1 \cdot x_2 \cdots x_n \leq \left(\frac{x_1 + x_2 + \cdots + x_n}{n}\right)^n, \quad x_1, x_2, \ldots, x_n > 0, n \geq 1. \tag{1.2}$$

Let $A_n = \frac{x_1+x_2+\cdots+x_n}{n}$. Since $\frac{A_n}{A_{n-1}} > 0$, taking $x := \frac{A_n}{A_{n-1}} - 1 > -1$ in (1.1), we have

$$\left(\frac{A_n}{A_{n-1}}\right)^n \geq 1 + n\left(\frac{A_n}{A_{n-1}} - 1\right) = \frac{nA_n - (n-1)A_{n-1}}{A_{n-1}} = \frac{x_n}{A_{n-1}}.$$

It follows that $A_n^n \geq x_n A_{n-1}^{n-1}$. This allows proving (1.2) by induction: for $n = 1$ the inequality is obvious, and assuming it holds for $n-1$ we obtain

$$A_n^n \geq x_n A_{n-1}^{n-1} \geq x_n(x_1 \cdots x_{n-1}) = x_1 \cdots x_n,$$

as claimed.

Using the obtained inequality, in turn, we will show[2] that for every positive number a and every integer n, there is a number b_n, denoted $\sqrt[n]{a}$ and termed the nth root of a, such that $b_n^n = a$ (how could you check that such a

[1] Written as $\sqrt[n]{x_1 \cdot x_2 \cdots x_n} \leq \frac{x_1+x_2+\cdots+x_n}{n}$, this becomes the well-known inequality between the arithmetic and geometric means. However, we do not want to use the notion of root at this point.

[2] Following Daniel Daners, Ulmer Seminare 2013, Notebook 18, Three Line Proofs.

number is uniquely determined?). To this end, let's consider the sequence given recursively by

$$x_1 = a, \qquad x_{k+1} = \frac{1}{n}\left((n-1)x_k + \frac{a}{x_k^{n-1}}\right). \tag{1.3}$$

This sequence is bounded from below by 0: all its members are positive, which is easy to check by induction. Also, because of (1.2), we have

$$x_{k+1}^n = \left(\frac{\overbrace{x_k + \cdots + x_k}^{(n-1)\text{terms}} + \frac{a}{x_k^{n-1}}}{n}\right)^n \geq a\,,$$

and this proves that

$$nx_{k+1} = \left((n-1) + \frac{a}{x_k^n}\right)x_k \leq nx_k,$$

that is, that the sequence is non-increasing. Hence, it has the limit $b_n := \lim_{k\to\infty} x_k$. Letting k tend to infinity in (1.3), we obtain

$$b_n = \frac{1}{n}\left((n-1)b_n + \frac{a}{b_n^{n-1}}\right).$$

Simple algebra now shows that $b_n^n = a$, completing the proof.

Let's have a closer look at this argument. Besides somewhat straightforward (though ingenious) calculations, it involves the following important step:

> any non-increasing sequence that is bounded from below has a limit.

As we shall see later (see Exercise 1.5), this sentence is a disguised statement that the set of real numbers is complete, without holes, full.

Is this completeness completely obvious? It seems to be: from our childhood we became accustomed to the fact that real numbers can be identified with points on a line (this was not at all obvious before R. Descartes, though), and the line does not have holes. We were also taught that real numbers are limits of sequences of rational numbers, and we think of π, for example, in a similar way. Therefore, we tend to think of the set of real numbers as a completion of the set of rational numbers: if there is any hole in the latter set, a real number fills this place.[3]

[3] A formal proof that real numbers fill the gaps in the set of rational numbers can be found in [34].

By the way, in the reasoning presented above we take it for granted that $\mathbb{Q}$, the set of rational numbers, is holey. Are there any grounds for such prejudice? Of course, there are. To explain, we know that some computations do not make sense in $\mathbb{Q}$. For example, in $\mathbb{Q}$ $\sqrt{2}$ is meaningless, that is, $\sqrt{2}$ is not a rational number,[4] and this has a bearing on our previous argument on the existence of $\sqrt[n]{a}$.

For, if a is rational number then, by induction, all elements of the sequence obtained from the recurrence (1.3) are rational also. The algebra remains the same, proving that the sequence does not increase and is bounded from below. As we have just recalled, the limit cannot be a rational number for $a = n = 2$ (and a great many other cases). Thus we have found a sequence of rational numbers that converges to an irrational number. If we were unaware of the existence of irrational numbers (for some, something that cannot be expressed as a fraction $\frac{m}{n}$ where m and n are integers is as strange as a pink elephant and does not resemble a number at all), we would be forced to say that

> not all sequences of rational numbers that are non-increasing and bounded from below converge.

This, however, means that $\mathbb{Q}$ is not a 'full' set; this set is not complete, for it has holes. A closer look at $\mathbb{Q}$ reveals that between any two distinct rational numbers there are infinitely many non-rational numbers. One could even say that there are more holes than there are non-holes ($\mathbb{Q}$ is countable, $\mathbb{R}$ and $\mathbb{R} \setminus \mathbb{Q}$ are uncountable). The most significant difference between $\mathbb{Q}$ and $\mathbb{R}$ is that the former has holes whereas the latter is complete.

[4] For, supposing that

$$\sqrt{2} = \frac{l}{m} = \frac{2^{l_2}3^{l_3}\cdots p^{l_p}}{2^{m_2}3^{m_3}\cdots q^{m_q}},$$

where l_2 is the number of times 2 shows up in the prime factorization of l, and so on, then by taking squares and multiplying both sides by m^2 we obtain

$$2^{2m_2+1}3^{2m_3}\cdots q^{2m_q} = 2^{2l_2}3^{2l_3}\cdots p^{2l_p}.$$

Note that on the left-hand side 2 is raised to an odd power, but on the right-hand side it is raised to an even power. Since this contradicts uniqueness of prime factorization, $\sqrt{2}$ cannot be rational.

1.3 Achilles

Some readers may dislike the previous section. Not all of us think taking roots is something they fancy doing. It is hard to argue against such an attitude, for it is possible to have a good living (in fact, be a billionaire) while being illiterate. But, in fact, completeness lies at the heart of something that for ages fascinated philosophers.

To see what I mean, let's recall Zeno of Elea, one of the most prominent students of Parmenides, who lived around 490 to 430 BCE. He is mostly known for his paradoxes, which were to substantiate his teacher's beliefs that plurality, change and motion in particular are but an illusion. Let's look at the apparently most famous and representative of these paradoxes: Achilles and the tortoise (see Figure 1.1). We all know that nobody is able to outrun the swift Achilles – and this is definitely impossible for a tortoise. But is the latter truly in a hopeless position? Suppose that initially the tortoise is at a distance $d > 0$ away from its pursuer, and that Achilles runs $\frac{1}{k}$ times faster than the little animal (where $k < 1$). The ill-matched competition begins – the tortoise tries to escape, and Achilles chases it. However, before Achilles catches the tortoise, he needs to come to the place where the tortoise had been at the beginning, and by that time the tortoise has moved a little (by a distance kd). Thus, Achilles faces a similar situation to the one he had initially: he needs to chase the tortoise who is at a distance kd away. Again, by the time Achilles comes to the place where tortoise had been this time, the tortoise has moved slightly away. This cycle will repeat infinitely, without end! And so, Achilles will never catch the tortoise. Quite a paradox!

Some may see this argument as pure sophistry. Many others (those who see that this argument cannot be easily refuted) may in fact start to doubt the world they see with their eyes is real. Such cases are known in history – for example, Georgias of Leontinoi, one of the philosophical followers of Zeno, became famous for his three nihilistic statements that can roughly be expressed as follows (see [40], p. 23): 1) there is nothing, 2) even if there were something, that something could not be apprehended, and 3) even if something were apprehended, this knowledge could neither be communicated nor understood by others. This leads to the following bold hypothesis:

Ignorance (of mathematics) is harmful.

Let's take neither of these roads, for both are disastrous. Zeno's paradoxes cannot be taken lightly or disregarded, because, as stated by W. Tatarkiewicz

Figure 1.1 Way to go, tortoise!

(see [40] p. 25), 'Zeno's paradoxes (...) were inspiring for and discussed by outstanding philosophers, including Bayle, Descartes, Leibniz, Kant, Hegel, Herbart, Hamilton, Mili, Renouvier, Bergson, and Russell.' Great minds have contemplated these matters, and we should appreciate the solution that came with the development of modern mathematical analysis and with the theory of convergence of infinite series in particular in the nineteenth century (after over two thousand years!).

Here is an explanation of the paradox. First of all, we note a gap in Zeno's reasoning: the fact that something takes place infinitely many times need not imply that it will take place for ever. More specifically, the sum of infinitely many terms need not be infinite. And that's the crux of the matter.

Let us take an even closer look at the paradox. Let t_0 be the time Achilles needs to reach the place where the tortoise was initially. As we have noted before, by that time the tortoise moves away by the distance dk. Thus the time needed for Achilles to cover the latter distance is kt_0. In that time, the tortoise moves $k^2 d$ away, and the time needed for Achilles to cover this distance is $k^2 t_0$, etcetera. Notice that 'etcetera' is not a scary word anymore,[5] because

$$t_\infty := \sum_{n=0}^{\infty} k^n t_0 < \infty;$$

it is at t_∞ that Achilles catches the tortoise.

[5] Unless you are afraid of Latin.

Let's work out the details: for any natural N, we have

$$\sum_{n=0}^{N} k^n t_0 = \frac{1-k^{N+1}}{1-k} t_0$$

(to see this, it suffices to multiply both sides by $1-k$ and do a little canceling), and the last expression converges to $\frac{t_0}{1-k}$ as $N \to \infty$. It does, we hasten to add, because k is smaller than 1, Achilles being faster than the tortoise. For $k > 1$, the sum on the right converges to infinity and Achilles turns out to be not so swift after all. By the way, for $k = 1$ he is not so swift either, but the formula above is different (can you provide it?).

It is perhaps worth looking at the yet more specific case of $k = \frac{1}{2}$ (Achilles is twice as fast as the tortoise) and $t_0 = 1$. Here, we are dealing with the sum of the infinite series

$$1 + \frac{1}{2} + \frac{1}{4} + \cdots .$$

Some might argue that this sum cannot be equal to 2, because it never 'reaches' 2. But even such notorious doubters can be somewhat convinced: beyond reasonable doubt, the finite sums

$$1 + \frac{1}{2} + \frac{1}{4} + \cdots + \frac{1}{2^N}, \qquad N \geq 1,$$

increase with N, and are bounded from above by 2. Zeno's paradox disappears if we agree that a non-decreasing sequence that is bounded from above has a limit; it is immaterial whether the limit here is 2; what is important is that the limit exists – the limit is the time when Achilles catches the tortoise. In other words, Zeno's paradox disappears if we agree that (see Exercise 1.4)

> time is a complete space, a space without holes.

If we reject this assumption, there need not be a time when Achilles catches the tortoise, and we will need to admit that reality does not agree with reason. As we see, completeness influences our living with no quarter. It is simply impossible to live without it.

1.4 Metric Spaces, Cauchy Sequences, Completeness

Let us come back to the subject of Section 1.1. What lies behind the fact that on a map of a county there is a point that lies directly above the point it describes?

The reason seems to be related to the notion of distance: there is a constant $q \in (0, 1)$ (termed the scale of a map) equal to the ratio of distances between points on a map and corresponding points in the area the map describes (this ratio is independent of the choice of these points). It is thus reasonable to start with the notion of distance.

To recall, a set X equipped with a function d (called a metric), mapping $X \times X$ to $\mathbb{R}^+$ and satisfying the following three conditions, is said to be a metric space:

(a) for all $x, y \in X$, equality $d(x, y) = 0$ holds if and only if $x = y$,
(b) for all $x, y \in X$, we have $d(x, y) = d(y, x)$,
(c) for all $x, y, z \in X$, we have $d(x, z) \leq d(x, y) + d(y, z)$.

Of course, $d(x, y)$ is interpreted as a distance between points x and y. With this interpretation, the conditions given above are plausible and agree with our intuition nicely: a distance between two points is zero if and only if these points coincide, distance to y measured from x is the same as the distance to x measured from y, and the way from x to z that leads through an intermediate point y cannot be shorter than the way that leads directly from x to z. Because of the last intuition, condition (c) is termed the triangle inequality and is best visualized if x, y and z are thought of as vertices of a triangle.

If we think now of the county we live in as a metric space X (with distance measured with a measuring ruler – even if this ruler is really long), then by placing a map of the county on a ground we define a transformation of X. In this transformation, to a point $x \in X$ we assign the x' lying directly below the point describing x on the map. Then

$$d(x', y') = qd(x, y), \tag{1.4}$$

where, as before, $q < 1$ is the map's scale.

As we shall see in the next chapter, this property, together with completeness of X, is a key to the property discussed in Section 1.1. For now, having the notion of distance defined, let us think of how this notion can be used to define complete spaces. To this end, let us come back to the example of Section 1.2. We have seen there that the set of rational numbers is not complete, and the argument for that was that there exists a non-increasing sequence of rational numbers that is bounded from below, and yet does not converge (to a rational number).

This idea is promising: why not, following Augustin Louis Cauchy, define complete spaces with the help of sequences? Why not detect holes by examining appropriate sequences? In an abstract metric space, however, we cannot work with non-decreasing or non-increasing sequences because in an abstract metric space more often than not there is no natural order. Hence, we need to have a good (gut) feeling what is an 'appropriate' sequence. Cauchy's brilliant idea was to define them as follows:

1.1 Definition A sequence $(x_n)_{n\geq 1}$ of elements of a metric space X is said to be a Cauchy sequence (or a fundamental sequence), if for any $\epsilon > 0$ there is an $n_0 \geq 0$ such that $d(x_n, x_m) < \epsilon$, as long as $n, m \geq n_0$.

In other words, for any ϵ one can throw away a finite number of elements of a Cauchy sequence in such a way that distances between any two of the remaining elements will be smaller than ϵ. Readers should convince themselves (at least intuitively) that non-increasing sequences of real numbers that are bounded from below are fundamental.

Let's see how fundamental sequences are related to convergent sequences. The latter are, to recall, defined as follows.

1.2 Definition A sequence $(x_n)_{n\geq 1}$ of elements of a metric space X is said to converge if there is an $x \in X$, said to be its limit, such that $\lim_{n\to\infty} d(x_n, x) = 0$; that is, for all $\epsilon > 0$ there is an n_0 such that $d(x_n, x) < \epsilon$ for $n \geq n_0$.

It is easy to check that any sequence that converges is fundamental. For, if x is its limit then, given $\epsilon > 0$ we can find n_0 such that $d(x_n, x) < \frac{\epsilon}{2}$ for all $n \geq n_0$. By the triangle inequality this implies, however, that $d(x_n, x_m) < \epsilon$ for all $n, m \geq n_0$, proving the claim.

Nevertheless, the converse is not true: there are Cauchy sequences that do not converge, and it is precisely the existence of such sequences that indicates the existence of holes in a metric space. The basic idea is that Cauchy sequences behave as if they were convergent. If we cannot find a limit of a Cauchy sequence, we suspect that the space we examine is holey.

To gain some more insight and to see a connection between fundamental sequences and holes in a metric space, let us think about a being that is living in the punched ball of Section 1.1. He/she knows our Euclidean metric but does not have a way of looking at the ball he/she lives in 'from outside' and thus discovering a hole. He/she only thinks that a point with all three coordinates equal to zero is not a point at all; for him/her it is a no-point.[6] We arrange things this way because we want him/her to detect the existence of the hole

[6] Ancient Greeks asked 'how can nothing be something?', and there are still many who cannot accept the existence of zero as a number. Or a person with zero morale. My dear referee of the Polish edition hastened, however, to recall S. J. Lec's aphorism which roughly translated goes as follows: 'When I reached the bottom, I heard knocking from below.'

from the inside, without looking from the outside. If he/she is a mathematician he/she can try to use a fundamental sequence. For instance, he/she can think of $(x_n)_{n\geq 1}$ given by $x_n = (\frac{1}{n}, \frac{1}{n}, \frac{1}{n})$, and argue as follows: 'given $\epsilon > 0$, I can throw away a finite number of elements of this sequence in such a way that for any x_n, x_m of the remaining elements the distances

$$d(x_n, x_m) = \sqrt{3}\left|\frac{1}{n} - \frac{1}{m}\right| < \frac{\sqrt{3}}{\min(m,n)}$$

between them are smaller than ϵ: it suffices to throw away all x_n with $n \leq \sqrt{3}\epsilon^{-1}$. Hence, this sequence seems to converge. However,' – he/she continues – 'I cannot think of a point that could be the limit of this sequence. For, any point, say x, in my decent space – the best space one can live in – has three coordinates of which at least one, say a, is non-zero (how ugly it would be for a point in my space to have all coordinates equal zero!). This shows, then, that the distance between x_n and x is at least $|\frac{1}{n} - a|$, and the latter quantity cannot converge to 0. I have thus found a Cauchy sequence that cannot converge. There must be something wrong with my space. I wonder what is it? Does it have a hole?'.

These considerations lead us to the following definition.

1.3 Definition A metric space is said to be complete if all fundamental (Cauchy) sequences of elements of this space converge.

As already discussed, the space of real numbers with distance $d(x, y) = |x - y|$ is a basic example of a complete space. On the other hand, the space of rational numbers, with the same distance, is full of holes. We will not give a formal proof of the fact that reals form a complete space – this is done in any decent course of real analysis (see e.g., [34]). Instead, we will show how completeness of $\mathbb{R}$ implies completeness of $\mathbb{R}^k, k \in \mathbb{N}$ when equipped with a Euclidean metric; more examples of complete spaces will be presented later in the book.

The argument that $\mathbb{R}^k$ is complete is in fact quite simple. For, let $(x_n)_{n\geq 1}$, where

$$x_n = (\xi_{n,1}, \xi_{n,2}, \ldots, \xi_{n,k}) \in \mathbb{R}^k$$

is a Cauchy sequence in $\mathbb{R}^k$. An easy-to-establish inequality

$$|\xi_{n,i} - \xi_{m,i}| \leq d(x_n, x_m) \tag{1.5}$$

(which holds for all $i = 1, \ldots, k$) implies that the numerical sequences $\left(\xi_{n,i}\right)_{n\geq 1}, i = 1, \ldots, k$ are fundamental in $\mathbb{R}$. Indeed, this inequality shows that $|\xi_{n,i} - \xi_{m,i}|$ is smaller than a given ϵ whenever $d(x_n, x_m)$ is smaller than this ϵ, and we know by assumption that $d(x_n, x_m)$ is smaller than an ϵ for all sufficiently large n, m. Since $\mathbb{R}$ is complete, there exist the limits $\xi_i = \lim_{n\to\infty} \xi_{n,i}, i = 1, \ldots, k$. Vector $x = (\xi_1, \xi_2, \ldots, \xi_k)$ is a member of $\mathbb{R}^k$. We are left with showing that x is a limit of $(x_n)_{n\geq 1}$. By assumption, for all $\epsilon > 0$ there is $n_0(\epsilon)$ such that

$$d(x_n, x_m) = \sqrt{\sum_{i=1}^{k} (\xi_{n,i} - \xi_{m,i})^2} < \epsilon$$

as long as $n, m \geq n_0(\epsilon)$. Letting $m \to \infty$, we obtain

$$d(x_n, x) = \sqrt{\sum_{i=1}^{k} (\xi_{n,i} - \xi_i)^2} \leq \epsilon$$

for $n \geq n_0(\epsilon)$. Hence, for $n \geq n_1(\epsilon) := n_0(\epsilon/2)$, we have $d(x_n, x) < \epsilon$, completing the proof.

1.5 Yet Another Encounter

As I have already mentioned above, completeness of metric spaces is a key to a number of theorems in pure and applied mathematics. Here is another, slightly more advanced, example: Dirichlet's test for convergence of functional series. The test says that a series of the form

$$\sum_{i=1}^{\infty} a_i x_i(s), \qquad s \in S,$$

where S is a set, a_i's are positive numbers and $x_i : S \to \mathbb{R}$ are functions, converges uniformly with respect to s, provided that the following two conditions are satisfied:

(a) $a_{i+1} \leq a_i$ for all $i \geq 1$ and $\lim_{i\to\infty} a_i = 0$,
(b) there is an $M > 0$ such that $|\sum_{i=1}^{n} x_i(s)| \leq M$ for all $s \in S$ and $n \geq 1$.

All one needs to know to prove validity of this test, besides a bit of algebra, is that the space of *bounded*[7] functions on S is a complete metric space when equipped with the distance

$$d(x, y) = \sup_{s\in S} |x(s) - y(s)|,$$

[7] Note that, were either x or y not bounded, $d(x, y)$ could be infinite.

and after reading a couple of chapters that follow, the reader will be able to check this completeness with ease, see Exercise 3.10.

As for the algebra, we first let

$$y_n(s) := \sum_{i=1}^{n} a_i x_i(s),$$

$$z_n(s) := \sum_{i=1}^{n} x_i(s), \qquad s \in S, n \geq 1.$$

By assumption (b), we have $|z_n(s)| \leq M$ for all $s \in S$ and $n \geq 1$. Then, as long as $m > n \geq 1$,

$$\begin{aligned} y_m(s) - y_n(s) &= \sum_{i=n+1}^{m} a_i[z_i(s) - z_{i-1}(s)] = \sum_{i=n+1}^{m} a_i z_i(s) - \sum_{i=n}^{m-1} a_{i+1} z_i(s) \\ &= \sum_{i=n+1}^{m-1} (a_i - a_{i+1}) z_i(s) + a_m z_m(s) - a_{n+1} z_n(s). \end{aligned}$$

Thus, since $a_i \geq a_{i+1}$, we see that $|y_m(s) - y_n(s)|$ does not exceed

$$M[a_m + \sum_{i=n+1}^{m-1} (a_i - a_{i+1}) + a_{n+1}] = 2Ma_{n+1},$$

yielding

$$d(y_m, y_n) \leq 2Ma_{n+1}.$$

Now, the second part of assumption (a) tells us that $(y_n)_{n\geq 1}$ is a Cauchy sequence. There is thus a bounded function y on S such that $\lim_{n\to\infty} d(y_n, y) = 0$, that is,

$$\sup_{s\in S} |y(s) - \sum_{i=1}^{n} a_i x_i(s)|$$

converges to 0, as $n \to \infty$. But this is precisely the uniform convergence of the series (to y).

1.6 Exercises

Exercise 1.1. Prove the Bernoulli inequality.

Exercise 1.2. Let $(x_n)_{n\geq 1}$ be a sequence of elements of a metric space, and suppose that $(x_n)_{n\geq 1}$ converges to an x in this space. Use the triangle inequality to show that then, for any y in this space, the numerical sequence $(d(x_n, y))_{n\geq 1}$ converges to $d(x, y)$.

Exercise 1.3. Check that $\sqrt{3}$ and $\sqrt{5}$ are not rational.

Exercise 1.4. Show that if any non-increasing sequence of reals that is bounded from below converges, then so does any any non-decreasing sequence of reals that is bounded from above.

Exercise 1.5. ⚠ Prove that completeness of the space of reals implies that any non-increasing sequence of reals that is bounded from below converges. Show also that the fact that any non-increasing sequence of reals that is bounded from below converges, implies completeness of the space of reals.

Exercise 1.6. Let $X := \{1, \frac{1}{2}, \frac{1}{2^2}, \frac{1}{2^3}, \cdots\} \subset \mathbb{R}$ be equipped with the metric $d(x, y) = |x - y|$. Is X complete? What about $X \cup \{0\}$?

Exercise 1.7. Prove (1.5). **Hint:** A sum of non-negative terms is no smaller than any of these terms.

Exercise 1.8. Let X be the space of sequences $(\xi_i)_{i\geq 1}$ such that ξ_i is either $+1$ or -1 for each $i \geq 1$. Check to see that this is a complete metric space with metric defined as follows:

$$d(x, y) := \sum_{i=1}^{\infty} \frac{1}{2^i} |\xi_i - \eta_i|,$$

where $x = (\xi_i)_{i\geq 1}$ and $y = (\eta_i)_{i\geq 1}$.

Exercise 1.9. If you are already convinced that Achilles will catch the tortoise, the time needed for him to do that can be found without summing the infinite series of Section 1.3, but simply by calculating t_∞ from the following relations:

$$k(d + x) = x, \qquad t_\infty = \frac{x}{v_{\text{tortoise}}}, \qquad t_0 = \frac{d}{v_{\text{Achilles}}} = \frac{dk}{v_{\text{tortoise}}}.$$

Provide the details.

Exercise 1.10. Argue as in Section 1.5 to prove the following test for convergence of series, attributed to Weierstrass (and known as the M-test). Suppose $a_n, n \geq 1$ are positive numbers such that $\sum_{n=1}^{\infty} a_n < \infty$ and $x_n, n \geq 1$ are functions on a set S. Assume that

$$|x_n(s)| \leq a_n, \qquad n \geq 1, s \in S.$$

Then the series $\sum_{n=1}^{\infty} x_n(s)$ converges absolutely and uniformly.

☞ CHAPTER SUMMARY

Guided by basic intuitions, we introduce the notion of a complete metric space and discover that we have in fact encountered it before in our study of mathematics. In particular, we learn that if the set of real numbers were not complete, bounded increasing (or decreasing) sequences would not have limits. Similarly, we realize that if time were not complete, Achilles would never catch the tortoise. In a slightly more advanced part, we show that the criteria for convergence of functional series involve the notion of completeness of the space of continuous functions.

2
Banach's Principle

2.1 Banach's Principle

This chapter is devoted to the first, quite simple and yet very elegant and useful, theorem in which completeness plays a crucial role: Banach's principle, also known as Banach's fixed point theorem.[1] In this theorem, 'geometric' intuitions presented in the foregoing chapter, where we discussed maps, are used, and presented in abstract terms. As we shall see later, such a generalization leads to a number of applications. In particular, we will be finally able to give a formal proof of the claim about maps made in Section 1.1.

Let's start from a simple corollary to the definition of a complete space.

2.1 Theorem *A closed (non-empty) subset of a complete metric space is a complete metric space, when equipped with the metric inherited from the original space.*

Proof To recall, a subset $Y \subset X$ of a metric space is said to be closed if and only if the limit of a sequence $(x_n)_{n\geq 1}$ of elements of Y belongs to Y. Note that for any $y_1, y_2 \in Y \subset X$ their distance is pre-defined in X; Y equipped with the related metric is also a metric space. This is what we have in mind when we say that Y inherits the metric from X.

We are to prove that each Cauchy sequence $(x_n)_{n\geq 1}$ of elements of Y converges (to an element of Y). However, from the perspective of X, $(x_n)_{n\geq 1}$ is a Cauchy sequence in the latter space, and so it converges to an $x \in X$. Using the closedness assumption, we see that $x \in Y$, completing the proof. □

[1] Banach himself would probably not be so happy knowing that such a simple theorem is attributed to his name: he has proved much deeper results (we will see some of them later). Nevertheless, even such a simple result shows how powerful the notion of completeness is.

2.2 Remark For those readers who are seeing □ for the first time: this is a standard graphic symbol used to mark the end of a proof. This end-of-proof sign will accompany us throughout the book.

2.3 Theorem (Banach's principle) *Suppose that a function T maps a complete mertic space into itself in such a way that there exists a constant $q \in (0, 1)$ such that, for all $x, y \in X$, we have*

$$d(Tx, Ty) \leq q\, d(x, y). \tag{2.1}$$

Then, there exists a unique fixed point $\tilde{x} \in X$ of T (this means that for this point we have $T\tilde{x} = \tilde{x}$, and there are no other points with this property); moreover, $\tilde{x}$ is the limit of $(T^n x)_{n\geq 1}$:

$$\tilde{x} = \lim_{n\to\infty} T^n x$$

regardless of the choice of $x \in X$.

($T^n x$ is defined inductively here: $T^{n+1}x = T(T^n x)$ and $T^1 x = Tx$.)

Proof Fix $x \in X$. First, we will show that $(T^n x)_{n\geq 1}$ converges. Since X is a complete space, it suffices to prove that $(T^n x)_{n\geq 1}$ is a Cauchy sequence. To this end, we will estimate the distance between $T^n x$ and $T^m x$ where $m > n$. Using the triangle inequality a number of times, we see that

$$d(T^n x, T^m x) \leq d(T^n x, T^{n+1}x) + d(T^{n+1}x, T^{n+2}x) + \cdots + d(T^{m-1}x, T^m x).$$

Assumption (2.1) used n times shows that the first term here does not exceed $q^n d(x, Tx)$. Similarly, the second does not exceed $q^{n+1}d(x, Tx)$, and so on. Hence,

$$\begin{aligned} d(T^n x, T^m x) &\leq d(x, Tx)[q^n + q^{n+1} + \cdots + q^{m-1}] \\ &= q^n d(x, Tx)[1 + q + q^2 + \cdots + q^{m-n-1}] \\ &\leq q^n d(x, Tx)[1 + q + q^2 + \cdots + q^{m-n-1} + \cdots] \\ &= \frac{q^n d(x, Tx)}{1 - q}. \end{aligned}$$

We note that the right-hand side does not depend on m anymore. This proves that $(T^n x)_{n\geq 1}$ is a Cauchy sequence, because the right-hand side converges to 0, as $n \to \infty$: if we want the distance between $T^n x$ and $T^m x$ to be small, it suffices to choose a sufficiently large n.

Next, let $\tilde{x}$ be the limit of $(T^n x)_{n\geq 1}$. Assumption (2.1) shows that T is continuous. It follows that $\lim_{n\to\infty} T(T^n x) = T(\tilde{x})$. On the other hand, $T(T^n x) = T^{n+1}x$, and the sequence $\left(T^{n+1}x\right)_{n\geq 1}$ obviously converges to the same limit as $(T^n x)_{n\geq 1}$. This proves that $\tilde{x} = \lim_{n\to\infty} T^{n+1}x = \lim_{n\to\infty} T(T^n x) = T(\tilde{x})$, and this means that $\tilde{x}$ is a fixed point of T. Uniqueness of $\tilde{x}$ is a direct consequence of (2.1): if there were two fixed points, say $\tilde{x}$ and $\hat{x}$, we would have

$$d(\tilde{x},\hat{x}) = d(T\tilde{x},T\hat{x}) \leq qd(\tilde{x},\hat{x}).$$

Since $q \in (0,1)$, this inequality forces $d(\tilde{x},\hat{x}) = 0$, implying $\tilde{x} = \hat{x}$. □

Maps that satisfy condition (2.1) are termed *contractions.*

As a direct consequence of Banach's principle we obtain the statement on maps made at the beginning of the previous chapter. For, as we have seen, each map lying in the area it depicts induces a map satisfying (1.4). Certainly, (1.4) is much stronger than (2.1). Thus, assumptions of Banach's principle are met ... as long as the area can be considered as a complete space. However, since this area can be identified with a subset of the plane,[2] by Theorem 2.1 it suffices to assume that the county or city we have in mind includes its borders, or, in other words, that it forms a closed set. Under these natural assumptions we infer that there is precisely one fixed point for the map in question. This, however, means that one and only one point lies precisely in the place it describes.

The next example seems to belong to linear algebra, but I am afraid without the notion of metric solving it would be much more difficult.

2.4 Example Suppose $\gamma_{i,j}, i, j = 1, \ldots, k$ are numbers such that

$$\sum_{j=1}^{k} |\gamma_{i,j}| < 1 \qquad \text{for all } i = 1, \ldots, k.$$

Then, for any $\eta_1, \ldots, \eta_k \in \mathbb{R}$, the system

$$\xi_j - \sum_{i=1}^{k} \gamma_{i,j}\xi_i = \eta_j \qquad \text{for all } j = 1, \ldots, k \tag{2.2}$$

has precisely one solution $x = (\xi_1, \ldots, \xi_k) \in \mathbb{R}^k$.

For X we take $\mathbb{R}^k$, equipped with the distance $d(x, \widetilde{x}) = \sum_{i=1}^{k} |\xi_i - \widetilde{\xi}_i|$, where $x = (\xi_1, \ldots, \xi_k)$ and $\widetilde{x} = (\widetilde{\xi}_1, \ldots, \widetilde{\xi}_k)$; as in Section 1.4 it can be checked that X is a complete metric space. Next, we let $T: X \to X$ be given by

[2] My valuable referee for the Polish edition of the book hastened to add: this is if we assume that the Earth is flat, which is disputable.

$$T(\xi_i)_{i=1,\dots,k} = \left(\eta_j + \sum_{i=1}^{k} \gamma_{i,j}\xi_i\right)_{j=1,\dots,k};$$

this is multiplying the matrix $(\gamma_{i,j})_{i,j=1,\dots k}$ by a row-vector from the left and adding the vector $(\eta_1, \dots, \eta_k)$. Then

$$\begin{aligned} d(Tx, T\widetilde{x}) &\leq \sum_{j=1}^{k}\sum_{i=1}^{k} |\gamma_{i,j}|\,|\xi_i - \widetilde{\xi}_i| = \sum_{i=1}^{k} |\xi_i - \widetilde{\xi}_i| \sum_{j=1}^{k} |\gamma_{i,j}| \\ &\leq \gamma d(x, \widetilde{x}), \end{aligned}$$

where

$$\gamma := \max_{i=1,\dots,k} \sum_{j=1}^{k} |\gamma_{i,j}|.$$

Since, by assumption, $\gamma < 1$, Banach's principle says that there is precisely one x such that $x = Tx$, and this means that there is precisely one x that solves (2.2). See also Example 13.2, much further on.

2.2 Exercises

Exercise 2.1. Given $y \in \mathbb{R}^k$, consider the map $x \mapsto x + y$. As long as $y \neq 0$, this map has no fixed points. With this example in mind answer the question of whether in Banach's principle the assumption that $q \in (0, 1)$ can be replaced by $q \in (0, 1]$.

Exercise 2.2. Show that the map $T\colon [1, \infty) \to [1, \infty)$ given by $f(x) = \frac{3}{4}(x + \frac{1}{x})$ is a contraction (assuming the standard distance in $[1, \infty)$). Characterize the fixed point of T.

Exercise 2.3. (Cf [21] p. 201). Let $\alpha \geq 1$ be given. Show that the equation

$$x^\alpha - (1 + \alpha)(1 - x) = 0$$

has precisely one solution in the interval $(0, 1)$. **Hint:** Consider the map $T\colon [0, 1] \to [0, 1]$ given by $T(x) = 1 - \frac{x^\alpha}{1+\alpha}$.

Exercise 2.4. Consider the space of Exercise 1.8, and the operator T that shifts each member of X by one coordinate to the right and completes the sequence by putting -1 on the first coordinate. In other words,

$$T(\xi_i)_{i\geq 1} = (-1, \xi_1, \xi_2, \dots).$$

Check to see that T satisfies (2.1) with $q = \frac{1}{2}$. Can you guess the fixed point of T?

Exercise 2.5. Suppose that, as in Example 2.4, $\gamma_{i,j}, i, j = 1, \ldots, k$ are numbers such that

$$\sum_{j=1}^{k} |\gamma_{i,j}| < 1 \qquad \text{for all } i = 1, \ldots, k.$$

Check that for any $\eta_1, \ldots, \eta_k \in \mathbb{R}$, the system

$$\xi_i - \sum_{j=1}^{k} \gamma_{i,j}\xi_j = \eta_i \qquad \text{for all } i = 1, \ldots, k \tag{2.3}$$

has precisely one solution $x = (\xi_1, \ldots, \xi_k) \in \mathbb{R}^k$. **Hint:** Argue as in Example 2.4, but equip $\mathbb{R}^k$ with the metric given by $d(x, \widetilde{x}) = \max_{i=1,\ldots,k} |\xi_i - \widetilde{\xi}_i|$.

Exercise 2.6. Let X be a metric space with metric d. Assume that T_n are maps $T_n \colon X \to X$ such that

$$d(T_n x, T_n y) \le q d(x, y), \qquad x, y \in X, n \ge 1,$$

for a certain $q \in (0, 1)$ that does not depend on n. Assume furthermore that all the limits

$$Tx := \lim_{n\to\infty} T_n x, \qquad x \in X,$$

exist. Prove that $T \colon X \to X$ has precisely one fixed point, say, $\tilde{x}$, and that the fixed points of the T_n's converge to this $\tilde{x}$.

☞ CHAPTER SUMMARY

Banach's principle states that if a map T uniformly reduces the distance between points of a complete metric space, then there is a unique $\widetilde{x}$ such that $T\widetilde{x} = \widetilde{x}$, called T's fixed point. This simple statement has profound and surprising consequences, as we will see in the following chapters. For now, we content ourselves with an example, which may appear to belong to the realm of linear algebra, but is, in fact, much easier to deal with using metric notions.

3
Picard's Theorem

As a first application of Banach's principle, and thus an application of the notion of completeness, in this chapter we discuss the Picard theorem on the existence and uniqueness of solutions to differential equations. The argument we present here is typical for functional analysis, which views functions (and solutions to differential equations in particular) as members of a certain space, and deduces the existence (or non-existence) of searched-for objects from properties, such as completeness, of this space. Hence, functional analysis is not concerned with properties of a particular function, such as monotonicity, continuity or integrability, but rather focuses on the properties of the entire space considered.

The main goal of this chapter is to point out that it is the completeness of the space of continuous functions that renders the existence and uniqueness of solutions to differential equations. Additionally, the reader will probably enjoy the elegant method of the weighted norm, due to Adam Bielecki, which allows one, in particular, to avoid the process of glueing together solutions defined in adjacent intervals and to obtain them in one simple move instead [10, 16].

3.1 Existence and Uniqueness of Solutions to Differential Equations

Let's consider an explicit first order differential equation, that is an equation of the form

$$u'(t) = f(t, u(t)), \qquad t \geq 0, \tag{3.1}$$

where u is a searched-for function, and f is given. If f is real-valued, we are looking for a real-valued u also. For example, for the Malthus equation, $f(t,u) = au$, and for the logistic equation, $f(t,u) = a\left(1 - \frac{u}{K}\right)u$, where a

and K are given constants. If $f: \mathbb{R}^+ \times \mathbb{R}^k \to \mathbb{R}^k$, where k is an integer, we are searching for a u with values being k-dimensional vectors. In other words, we are searching for k real-valued functions $u_1, u_2, \ldots, u_k$ forming u, as its coordinates:

$$u(t) = (u_1(t), u_2(t), \ldots, u_k(t)).$$

For instance, in the case of the so-called Lotka–Volterra equation we are searching for a function $u = (u_1, u_2)$ with values in $\mathbb{R}^2$; the function f is given by

$$f(t, u_1, u_2) = (au_1 - bu_1u_2, \ -cu_2 + du_1u_2) \tag{3.2}$$

and maps $\mathbb{R}^+ \times \mathbb{R}^2$ to $\mathbb{R}^2$, that is, it assigns a pair of numbers to a triple of numbers; a, b, c and d are positive constants. (As in all the foregoing examples, the value of f given here does not depend on its first argument, interpreted as time. This means that we are facing a so-called *autonomous* equation. In a moment, we will see an example of a non-autonomous equation.)

One of the fundamental theorems of the theory of ordinary differential equations, the theorem of Peano, says that if f is continuous in a neighborhood of $(0, \alpha) \in \mathbb{R}^{k+1}$, then in a certain small neighborhood of $t = 0$ there exists a solution of (3.1) satisfying the initial condition $u(0) = \alpha$. (Such a solution is known as the solution to the initial value problem or Cauchy problem.) We stress that the latter neighborhood may be quite small and that its size depends on f and α. For example, the solution to the equation

$$u'(t) = [u(t)]^2 t^3 \tag{3.3}$$

with initial condition $u(0) = \alpha$ is given by

$$u(t) = \frac{4\alpha}{4 - \alpha t^4}.$$

For negative α, this u is defined on the entire real line, but for positive α the denominator equals 0 for $t = \sqrt[4]{\frac{4}{\alpha}}$ and thus u is ill-defined at this point. Hence, u is defined merely in the interval $[0, \sqrt[4]{\frac{4}{\alpha}})$ (see Figure 3.1).

It should be stressed that Peano's theorem is concerned with the existence of local solutions and says nothing about their uniqueness. Thus it may happen that there are many solutions to the same equation going out from a single point. This is the case, for example, for the function $f(t, u) = 2\sqrt{u}$. The related initial value problem,

$$u'(t) = 2\sqrt{|u(t)|}, \qquad u(0) = 0, \tag{3.4}$$

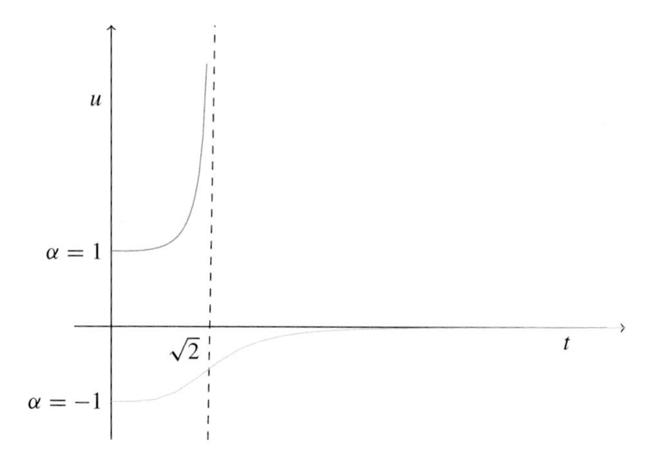

Figure 3.1 Local solutions to ordinary differential equations: the upper graph is the solution to equation (3.3) with $\alpha = 1$; the lower graph is the solution with $\alpha = -1$.

has two distinct solutions: $u(t) = 0$ and $u(t) = t^2$. In fact, there are infinitely many solutions to this problem, indexed by $a > 0$. To see this, given $a > 0$, define u_a by

$$u_a(t) = \begin{cases} 0, & \text{for } t \in [0, a], \\ (t-a)^2, & \text{for } t > a. \end{cases}$$

Then u_a satisfies (3.4) and clearly, $u_a \not\equiv u_b$ for $a \neq b$. Similarly, the problem

$$u'(t) = \sqrt[3]{u(t)}, \qquad u\,(0) = 0$$

has three solutions: $u(t) = 0, u(t) = -\left(\frac{2}{3}t\right)^{\frac{3}{2}}$ and $u(t) = \left(\frac{2}{3}t\right)^{\frac{3}{2}}$, and infinitely many other solutions that can be obtained by glueing these three together.

Here is a classical condition that guarantees uniqueness of solutions: the famous Picard's theorem says that if f is continuous and there is a non-negative constant L ('L' for 'Lipschitz') such that

$$\|f(t,u) - f(t,\tilde{u})\| \leq L\|u - \tilde{u}\| \tag{3.5}$$

for all u and $\tilde{u}$ in $\mathbb{R}^k$, then for any initial value α there is a solution to the related differential equation; moreover, the solution is unique and defined on the entire real line. Here, $\|\cdot\|$ denotes the Euclidean norm,

$$\|(\xi_1, \ldots, \xi_k)\| := \sqrt{\xi_1^2 + \xi_2^2 + \cdots + \xi_k^2},$$

or any other, equivalent norm, for example the maximum norm:

$$\|(\xi_1,\ldots,\xi_k)\|_{\max} := \max\{|\xi_1|,|\xi_2|,\ldots,|\xi_k|\}.$$

(Since this, by assumption, is not a first course in functional analysis, the reader will forgive me for using the notion of the norm, which will be formally defined later.)

In the case of equation (3.4) we have a number of distinct solutions because the function $f(t,u) = \sqrt{u}$, even though it is continuous, is not Lipschitz continuous. That is, even if we restrict ourselves to $u \in [0,1]$, there is no L such that (see Exercise 3.3)

$$\sqrt{u} = \|f(t,u) - f(t,0)\| \le L\|u-0\| = Lu. \tag{3.6}$$

Is the Lotka–Volterra equation related to a Lipchitz continuous function? Yes and no. To explain this strange answer we need to be more precise. Namely, we will say that a function f satisfying (3.5) is *globally Lipchitz continuous* (with respect to u). It turns out that the function f of equation (3.2) is not globally Lipschitz continuous but it is locally Lipschitz continuous. This means that for each $r > 0$ there is $L(r)$ such that

$$\|f(t,u) - f(t,\tilde{u})\| \le L(r)\|u - \tilde{u}\|, \tag{3.7}$$

as long as the norms of u and $\tilde{u}$ do not exceed r. Indeed (see Exercise 3.6), condition (3.7) holds for such u and $\tilde{u}$ with $L(r) = \max\{a+2br, c+2dr\}$. We note also that this situation is quite typical in that $L(r)$ increases with r.

Unfortunately, local Lipchitz continuity, although it does guarantee the local existence and uniqueness of solutions, does not guarantee that these solutions are defined for all $t \ge 0$. This is clear from the example

$$u'(t) = 1 + [u(t)]^2.$$

As it is easy to check (see Exercise 3.4), the function $f(u) = 1 + u^2$ is locally Lipschitz continuous but not globally Lipschitz continuous. All solutions to this equation are of the form $u(t) = \tan(t + C)$, where C is a constant, and are defined on appropriate domains. In particular, the solution starting at $\alpha = 0$ is given by $u(t) = \tan t$ and obviously is defined merely for $t \in [0, \frac{\pi}{2})$.

3.2 The Space of Continuous Functions

It is completeness of the space of continuous functions that is the key to the proof of Picard's theorem. Therefore, in this section, we take a closer look at this space.

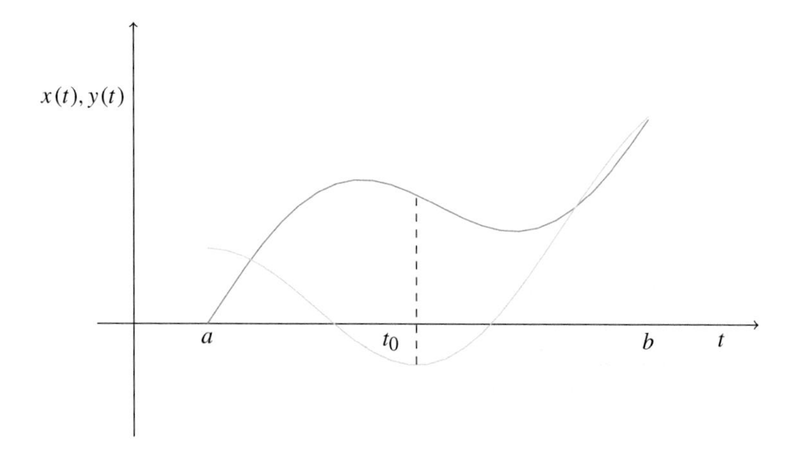

Figure 3.2 The supremum metric: the distance between two functions is the maximal length of a segment, perpendicular to the t axis, that connects their graphs. On the picture it is the length of the segment drawn with a dashed line.

Let $a < b$ be given numbers, and let $C[a,b]$ denote the space of all continuous functions on the interval $[a,b]$, taking values in $\mathbb{R}^k$. (If we were to be more formal, we should write $C([a,b],\mathbb{R}^k)$, but for simplicity we will use the notation given above.) Let's choose two functions, say x and y, from this set. The function ϕ given by $\phi(t) = \|x(t) - y(t)\|$, where $\|\cdot\|$ is the norm in $\mathbb{R}^k$, is then real-valued and continuous. According to a well-known result of real analysis, since the interval $[a,b]$ is compact, there is a $t_0 \in [a,b]$ where ϕ attains its maximum (see Figure 3.2). Hence, we can define the distance between x and y as follows:

$$d(x,y) = \max_{t\in[a,b]} \|x(t) - y(t)\| = \|x(t_0) - y(t_0)\|. \tag{3.8}$$

We stress that t_0 varies with x and y.

It is easy to check (see Exercise 3.8) that d so defined is a metric. We will check that $C[a,b]$ when equipped with this metric is a complete space. To this end, let's think of a Cauchy sequence $(x_n)_{n\geq 1}$ of elements of $C[a,b]$: we need to show that this sequence converges. To begin with, we will show that, for every $s \in [a,b]$, the sequence $(x_n(s))_{n\geq 1}$ converges. By assumption, we know that for any $\epsilon > 0$ there is an n_0 such that for $n,m \geq n_0$ we have

$$\max_{t\in[a,b]} \|x_n(t) - x_m(t)\| < \epsilon. \tag{3.9}$$

Since

$$\|x_n(s) - x_m(s)\| \le \max_{t\in[a,b]} \|x_n(t) - x_m(t)\|,$$

it follows that $(x_n(s))_{n\ge1}$ is a Cauchy sequence. But $(x_n(s))_{n\ge1}$ has values in $\mathbb{R}^k$, and the latter space is complete. Therefore, there exists $x(s) \in \mathbb{R}^k$ such that $\lim_{n\to\infty} x_n(s) = x(s)$.

The function x, assigning the limit $x(s) = \lim_{n\to\infty} x_n(s)$ to an $s \in [a,b]$, is a natural candidate for the limit of the sequence $(x_n)_{n\ge1}$. However, we need to show that x is a member of the space $C[a,b]$, that is, that x is a continuous function, and that $(x_n)_{n\ge1}$ converges to x in the sense of the metric d. Condition (3.9) tells us that if $n, m \ge n_0$ then, regardless of the choice of t, we have

$$\|x_n(t) - x_m(t)\| < \epsilon. \tag{3.10}$$

Letting $m \to \infty$, we see that, regardless of what $t \in [a,b]$ is, we have

$$\|x_n(t) - x(t)\| \le \epsilon, \tag{3.11}$$

as long as $n \ge n_0$. If we knew that x was continuous we could thus infer that $\lim_{n\to\infty} d(x_n, x) = 0$.

Let's then prove the continuity. To this end, we fix $t_0 \in [a,b]$ and $\epsilon > 0$. Next, we choose n_0 so that for $n \ge n_0$ the inequality (3.11) with ϵ replaced by $\frac{\epsilon}{3}$ holds. Since the function x_{n_0} is continuous (at t_0 in particular), we can choose a $\delta > 0$ so that conditions $t \in [a,b]$ and $|t - t_0| < \delta$ imply that $\|x_{n_0}(t) - x_{n_0}(t_0)\| < \frac{\epsilon}{3}$. For such t we can write

$$\begin{aligned}\|x(t) - x(t_0)\| &\le \|x(t) - x_{n_0}(t)\| + \|x_{n_0}(t) - x_{n_0}(t_0)\| + \|x_{n_0}(t_0) - x(t_0)\| \\ &< \frac{\epsilon}{3} + \frac{\epsilon}{3} + \frac{\epsilon}{3} = \epsilon.\end{aligned}$$

Since ϵ and t_0 are arbitrary, this shows that x is continuous and completes the entire proof.

We recall that the d introduced above is termed the maximum metric and that a sequence converging in the sense of this metric is said to converge uniformly in the interval $[a,b]$.

3.3 Equivalent Metrics

The argument used to prove the second (true) version of the local theorem of Picard (i.e., Theorem 3.3) involves also the notion of equivalent metrics.

Two metrics, d_1 and d_2, defined in the same space X are said to be (Lipschitz) equivalent if there are positive constants m and M such that

$$md_1(x, y) \leq d_2(x, y) \leq Md_1(x, y)$$

for all $x, y \in X$. We will need the following simple observation.

3.1 Theorem *If X is complete when equipped with metric d_1, and d_2 is (Lipschitz) equivalent to d_1, then X equipped with d_2 is complete also.*

Proof Think of a sequence $(x_n)_{n\geq 1}$ that satisfies the Cauchy condition in the metric d_2. The left-hand side of the definition of equivalent metrics implies that this sequence satisfies the Cauchy condition in the metric d_1, too (because $d_1(x_n, x_k) \leq \frac{1}{m} d_2(x_n, x_k)$). By assumption, therefore, there is an $x \in X$ such that

$$\lim_{n\to\infty} d_1(x_n, x) = 0.$$

Using the right-hand side of the definition, combined with the squeeze theorem for sequences, we infer then that

$$\lim_{n\to\infty} d_2(x_n, x) = 0,$$

completing the proof. □

Here is an example of equivalent metrics in the space $C[a,b]$. Let ψ be an arbitrary continuous function on $[a,b]$ with positive values. For $x, y \in C[a,b]$ we consider

$$d_\psi(x, y) = \max_{t\in[a,b]} \psi(t)\|x(t) - y(t)\|. \tag{3.12}$$

Arguing as in the case of (3.8), we check that this maximum exists and is finite. It is also easy to check that d_ψ is a metric (see Exercise 3.9). We claim that this metric is equivalent to (3.8). Indeed, ψ being continuous and having positive values, there are positive constants m and M such that

$$m \leq \psi(t) \leq M$$

for all $t \in [a,b]$. Therefore,

$$\begin{aligned} md(x, y) &= \sup_{t\in[a,b]} m\|x(t) - y(t)\| \\ &\leq \sup_{t\in[a,b]} \psi(t)\|x(t) - y(t)\| \quad (= d_\psi(x, y)) \\ &\leq \sup_{t\in[a,b]} M\|x(t) - y(t)\| = Md(x, y) \end{aligned}$$

for any $x, y \in C[a,b]$, proving the claim. In particular, by the theorem proved above, $C[a,b]$ is complete when equipped with d_ψ.

3.4 Local Theorem of Picard

We are now ready to prove the Picard theorem; our argument will be based on the Banach principle. We begin with the local version of the theorem.

3.2 Theorem (Local theorem of Picard I) *Let $a, b > 0$, $t_0 \in \mathbb{R}$ and $\alpha \in \mathbb{R}^k$ be given, and let f be a function with values in $\mathbb{R}^k$ defined on the 'rectangle'*

$$R := \{(t,x) \in \mathbb{R}^{k+1};\ |t - t_0| \leq a\,, \|x - \alpha\| \leq b\}.$$

Assume that f is continuous on R and Lipchitz continuous with respect to the second argument with constant L, that is,

$$\|f(t,x) - f(t,y)\| \leq L\|x - y\|$$

for all $(t,x), (t,y) \in R$. Moreover, let $M := \sup_{(t,x)\in R} \|f(t,x)\|$ (M is finite, because f is continuous on a compact set) and

$$\kappa := \min\left(a, \frac{b}{M}, \frac{1}{2L}\right). \tag{3.13}$$

Then, there exists precisely one solution to the problem

$$u'(t) = f(t, u(t)), \qquad u(t_0) = \alpha, \tag{3.14}$$

defined in the interval where $|t - t_0| \leq \kappa$.

Proof Our proof consists of three steps. In the first, we describe the space in which, in step three, the solution will be found as a fixed point of a certain transformation. In the second, intermediate, step we establish a connection between the differential equation and the transformation just mentioned.

1. As established in Section 3.2, $C[t_0 - \kappa, t_0 + \kappa]$ with appropriate distance is a complete space. Its subset X of functions that have values in the closed ball

$$\overline{B(\alpha,b)} := \{x \in \mathbb{R}^k;\ \|x - \alpha\| \leq b\}$$

is closed (see Exercise 3.2). Thus, by Theorem 2.1, X itself is a complete metric space.

2. Let's rewrite (3.14) in the equivalent *integral form*:

$$u(t) = \alpha + \int_{t_0}^{t} f(s, u(s))\,\mathrm{d}s. \tag{3.15}$$

The second term here is an integral of a vector-valued function: To recall, a map $g\colon [t_0-\kappa, t_0+\kappa] \to \mathbb{R}^k$ can be identified with k maps $g_i : [t_0-\kappa, t_0+\kappa] \to \mathbb{R}$, $i = 1, \dots, k$, so that $g(t) = (g_1(t), \dots, g_k(t))$ and then, by definition,

$$\int_{t_0}^{t} g(t)\,\mathrm{d}t = \left(\int_{t_0}^{t} g_1(t)\,\mathrm{d}t, \dots, \int_{t_0}^{t} g_k(t)\,\mathrm{d}t\right). \tag{3.16}$$

We note that this form includes the information about initial condition $u(t_0) = \alpha$. Nevertheless, its main advantage is that it allows us to reduce the problem of existence of solutions to a differential equation to the problem of existence of fixed points of a certain transformation.

3. We consider the transformation $T: X \to X$ given by $x \mapsto Tx$, where

$$(Tx)(t) = \alpha + \int_{t_0}^{t} f(s, x(s))\,\mathrm{d}s, \qquad t \in [t_0 - \kappa, t_0 + \kappa]. \tag{3.17}$$

The integral on the right-hand side is well defined because the assumption that $x \in X$ implies that $(s, x(s))$ belongs to R. We note that for $t \le t_0$ the integral is defined as $-\int_t^{t_0}$. We check that T does indeed transform X into X: first of all, Tx is a continuous function, and, second,

$$\begin{aligned}\|(Tx)(t) - \alpha\| &= \left\|\int_{t_0}^{t} f(s, x(s))\,\mathrm{d}s\right\| \le \left|\int_{t_0}^{t} \|f(s, x(s))\|\,\mathrm{d}s\right| \\ &\le M|t - t_0| \le M\kappa \le b,\end{aligned}$$

where the second inequality is a result of Exercise 3.13. Moreover,

$$\begin{aligned}d(Tx, Ty) &= \sup_{t\in[t_0-\kappa, t_0+\kappa]} \left\|\int_{t_0}^{t} [f(s, x(s)) - f(s, y(s))]\,\mathrm{d}s\right\| \\ &\le \sup_{t\in[t_0-\kappa, t_0+\kappa]} \left|\int_{t_0}^{t} \|f(s, x(s)) - f(s, y(s))\|\,\mathrm{d}s\right| \\ &\le L \sup_{t\in[t_0-\kappa, t_0+\kappa]} \left|\int_{t_0}^{t} \|x(s) - y(s)\|\,\mathrm{d}s\right| \\ &\le L \sup_{t\in[t_0-\kappa, t_0+\kappa]} \left|\int_{t_0}^{t} d(x, y)\,\mathrm{d}s\right| \\ &\le Ld(x, y) \sup_{t\in[t_0-\kappa, t_0+\kappa]} |t - t_0| \\ &= L\kappa\, d(x, y) \\ &\le \frac{1}{2} d(x, y).\end{aligned}$$

This calculation shows that T satisfies the assumptions of Banach's principle. Therefore, there is precisely one function u that is a fixed point for T. That is, there is one and only one function u such that (3.15) holds for all t in the interval $[t_0-\kappa, t_0+\kappa]$. This means, however, that u solves problem (3.14). But, any solution to (3.14) is also a solution to (3.15). This completes the proof. □

A closer inspection of the proof reveals that one of the assumptions in our theorem can be relaxed – in the definition of κ, instead of $\frac{1}{2L}$ one can take qL^{-1} where $q \in (0,1)$ and the argument will still work. In fact, in the 'true' Picard theorem, the value of the Lipschitz constant plays no role in defining κ, for we have

$$\kappa := \min\left(a, \frac{b}{M}\right). \tag{3.18}$$

But our calculations presented above do not lead to this result. For the full proof we need an equivalent metric in $C[a,b]$ (see [21], p. 203).

3.3 Theorem (Local theorem of Picard II) *Under the assumptions of Theorem 3.2, there is precisely one solution to the problem* (3.14) *that is defined in the interval where* $|t-t_0| \le \kappa$ *with* κ *given by* (3.18).

Proof Fix $\lambda > 0$, and let $\psi(t) = \mathrm{e}^{-\lambda|t-t_0|}, t \in [t_0-\kappa, t_0+\kappa]$. The ψ so defined is a continuous function with positive values. Thus, as we know from the previous subsection, the formula

$$d_\psi(x,y) = \sup_{t\in[t_0-\kappa,t_0+\kappa]} \mathrm{e}^{-\lambda|t-t_0|}\|x(t)-y(t)\|$$

defines a metric that makes $C[a,b]$, and its subset X described in the proof of Theorem 3.2, complete metric spaces. This metric is termed a Bielecki-type metric to honor Adam Bielecki [10, 16], who invented it.[1]

As in our first argument, we define X as the set of functions that have values in the closed ball with center at α and radius b, and we define the transformation T as in (3.17); we already know that T maps X into X. Moreover,

$$\begin{aligned}
d_\psi(Tx,Ty) &= \sup_{t\in[t_0-\kappa,t_0+\kappa]} \mathrm{e}^{-\lambda|t-t_0|}\left\|\int_{t_0}^{t}[f(s,x(s))-f(s,y(s))]\,\mathrm{d}s\right\| \\
&\le \sup_{t\in[t_0-\kappa,t_0+\kappa]} \mathrm{e}^{-\lambda|t-t_0|}\left|\int_{t_0}^{t}\|f(s,x(s))-f(s,y(s))\|\,\mathrm{d}s\right| \\
&\le L \sup_{t\in[t_0-\kappa,t_0+\kappa]} \left|\int_{t_0}^{t}\mathrm{e}^{-\lambda[|t-t_0|-|t_0-s|]}\{\mathrm{e}^{-\lambda|t_0-s|}\|x(s)-y(s)\|\}\,\mathrm{d}s\right|.
\end{aligned}$$

[1] Adam Bielecki is probably better known for his work on the axioms of Euclid, and he himself definitely valued this work more than the invention of the distance discussed here.

To proceed, we note that, regardless of whether t is smaller or greater than t_0, the variable s lies between these two points. Hence, $|t - t_0| - |t_0 - s| = |t - s|$. Furthermore, the expression in brackets does not exceed $d_\psi(x, y)$. Therefore,

$$\begin{aligned} d_\psi(Tx, Ty) &\leq L \sup_{t \in [t_0 - \kappa, t_0 + \kappa]} \left| \int_{t_0}^{t} e^{-\lambda|t-s|} d_\psi(x, y)\, ds \right| \\ &= L d_\psi(x, y) \sup_{t \in [t_0 - \kappa, t_0 + \kappa]} \int_0^{|t - t_0|} e^{-\lambda u}\, du \\ &= L \int_0^{\kappa} e^{-\lambda u}\, du \; d_\psi(x, y) < \frac{L}{\lambda} d_\psi(x, y). \end{aligned}$$

This shows that for $\lambda > L$, T is a contraction in X with respect to the norm d_ψ. In other words, Banach's principle is in force and the argument can be carried out as in the previous proof. □

3.5 Global Theorem of Picard

In our final section we will prove the global version of Picard's theorem. The local version is concerned with solutions that are defined in a small neighborhood of the point where the initial condition is prescribed. In the global version, under a slightly stronger assumption, we will be able to deduce the existence of unique solutions defined on the entire time half-line.

3.4 Theorem (Global theorem of Picard) *Assume that f is continuous, and that it is globally Lipschitz (i.e., that (3.5) is satisfied). Assume furthermore that there are constants M_0 and ω_0 such that*

$$\left\| \int_0^t f(s, 0)\, ds \right\| \leq M_0 e^{\omega_0 t}, \qquad t \geq 0. \tag{3.19}$$

Then for all $\alpha \in \mathbb{R}^k$, equation (3.1) has precisely one solution satisfying the initial condition $u(0) = \alpha$. This solution is defined for all $t \geq 0$ and grows at most exponentially, that is, there are constants M and $\omega \geq \omega_0$ such that

$$\|u(t)\| \leq M e^{\omega t}, \qquad t \geq 0.$$

3.5 Remark As before (see (3.16)), the integral on the left-hand side of (3.19) is an integral of a vector-valued function, and thus is obtained by integrating coordinate by coordinate. A condition of type (3.19) is necessary if we want to obtain solutions that grow at most exponentially. To see this consider $f(t, u) = 2te^{t^2}$; the function so defined is globally Lipschitz continuous with respect to u, with $L = 0$, and solutions to the corresponding differential equation have the form $u(t) = \alpha e^{t^2}$ and thus grow faster than exponentially. Condition (3.19) is automatically satisfied for autonomous differential equations, that is in the case where f does not depend on t.

Proof of the global theorem of Picard We argue similarly as in the proof of the local theorem. The main difference is that the space where the solution will be found is a little more complicated than before.

(a) Given $\omega > 0$, let $C_\omega(\mathbb{R}^+)$ be the space of continuous functions $x\colon [0,\infty) \to \mathbb{R}^k$ such that there is a finite $M = M(x)$ satisfying

$$\sup_{t\geq 0} e^{-\omega t}\|x(t)\| \leq M.$$

In other words, members of $C_\omega(\mathbb{R}^+)$ are continuous functions of exponential growth (with exponent ω). Our first goal is to show that this space with Bielecki distance

$$d_\omega(x,y) = \sup_{t\geq 0} e^{-\omega t}\|x(t) - y(t)\|$$

is a complete metric space. The reader is asked to check that d_ω is a well-defined metric (see Exercise 3.11); we will restrict ourselves to the more challenging task of proving the completeness of $C_\omega(\mathbb{R}^+)$.

Let $(x_n)_{n\geq 1}$ be a Cauchy sequence in $C_\omega(\mathbb{R}^+)$. The inequality

$$\sup_{t\in[0,\tau]} \|x_n(t) - x_m(t)\| \leq e^{\omega\tau} d_\omega(x_n,x_m), \tag{3.20}$$

which holds for all $\tau > 0$, shows that for all t there exists the limit $x(t) := \lim_{n\to\infty} x_n(t)$. Moreover, functions x_n converge uniformly to the limit function in every interval $[0,\tau]$, as this inequality shows that $x_n, n \geq 0$, when restricted to an interval $[0,\tau]$, form a Cauchy sequence in the space $C[0,\tau]$.[2] In particular, x is continuous on every interval $[0,\tau]$ (see the proof of completeness of the space $C[a,b]$), and so is continuous on the entire half line where $t \geq 0$.

We still need to check that x belongs to $C_\omega(\mathbb{R}^+)$ and that $(x_n)_{n\geq 1}$ converges to x in the sense of the metric d_ω. Since $(x_n)_{n\geq 1}$ is a fundamental sequence, for each $\epsilon > 0$ there is an n_0 such that

$$\text{for all } t \geq 0,\ \ e^{-\omega t}\|x_n(t) - x_m(t)\| < \epsilon.$$

[2] Let's be more specific about this reasoning: given $\epsilon > 0$, and having fixed τ beforehand, we can choose n_0 so that $d_\omega(x_n,x_m) < \epsilon e^{-\omega\tau}$ for $n,m \geq n_0$. Relation (3.20) implies then that for such n and m we have $\sup_{t\in[0,\tau]} \|x_n(t) - x_m(t)\| < \epsilon$. Since ϵ is arbitrary, it follows that restrictions of x_n's to the interval $[0,\tau]$ form a Cauchy sequence.

Letting $m \to \infty$, we see that

$$\text{for all } t \geq 0, \ \mathrm{e}^{-\omega t} \|x_n(t) - x(t)\| \leq \epsilon, \tag{3.21}$$

as long as $n \geq n_0$. In particular, choosing M so that $\|x_{n_0}(t)\| \leq M\mathrm{e}^{\omega t}$, we obtain $\|x(t)\| \leq \epsilon \mathrm{e}^{\omega t} + M\mathrm{e}^{\omega t} \leq (M + \epsilon)\mathrm{e}^{\omega t}$. This means that x is of at most exponential growth with exponent ω, that is, $x \in C_\omega(\mathbb{R}^+)$. Inequality (3.21) shows furthermore that $d_\omega(x_n, x) \leq \epsilon$ as long as $n \geq n_0$, and this, ϵ being arbitrary, proves that $(x_n)_{n \geq 1}$ converges to x.

(b) Now, let ω be larger than the Lipschitz constant of (3.5) and larger than the ω_0 of the assumption (3.19). In the space $C_\omega(\mathbb{R}^+)$ we consider the map, say T, which to a function x assigns Tx given by

$$(Tx)(t) = \alpha + \int_0^t f(s, x(s)) \,\mathrm{d}s. \tag{3.22}$$

Function Tx is obviously continuous, but it is not yet clear whether it belongs to $C_\omega(\mathbb{R}^+)$.

Let's think, however, of two members, x and y, of $C_\omega(\mathbb{R}^+)$ and given $t \geq 0$, let's estimate the quantity

$$\begin{aligned}
\mathrm{e}^{-\omega t} \|(Tx)(t) - (Ty)(t)\| &= \mathrm{e}^{-\omega t} \| \int_0^t f(s, x(s)) \,\mathrm{d}s - \int_0^t f(s, y(s)) \,\mathrm{d}s \| \\
&\leq \mathrm{e}^{-\omega t} \int_0^t \|f(s, x(s)) - f(s, y(s))\| \,\mathrm{d}s \\
&\leq \mathrm{e}^{-\omega t} \int_0^t L\|x(s) - y(s)\| \,\mathrm{d}s;
\end{aligned}$$

the first inequality above is valid by Exercise 3.13. We proceed by estimating as follows:

$$\begin{aligned}
\mathrm{e}^{-\omega t} \|(Tx)(t) - (Ty)(t)\| &\leq \int_0^t L\mathrm{e}^{-\omega(t-s)} \mathrm{e}^{-\omega s} \|x(s) - y(s)\| \,\mathrm{d}s \\
&\leq \int_0^t L\mathrm{e}^{-\omega s} \,\mathrm{d}s \, d_\omega(x, y) \\
&< \frac{L}{\omega} d_\omega(x, y).
\end{aligned} \tag{3.23}$$

However, assumption (3.19) implies that for $y = 0$, Ty belongs to $C_\omega(\mathbb{R}^+)$. This combined with inequality (3.23) shows that Tx is a member of $C_\omega(\mathbb{R}^+)$ for all $x \in C_\omega(\mathbb{R}^+)$. The latter inequality shows also that T satisfies assumption of Banach's principle with $q = \frac{L}{\omega} < 1$. Therefore, there is precisely one fixed point of T in $C_\omega(\mathbb{R}^+)$, say u. This u is a searched-for solution to the differential equation, as u satisfies (3.15), and vice versa a solution to (3.1) must satisfy (3.15), that is, it must be a fixed point of T. The fact that the so-obtained u is of at most exponential growth is a bonus from our proof: we have found u in $C_\omega(\mathbb{R}^+)$. □

A final clarification is in order here. Our argument shows that there are no global exponentially growing solutions to the differential equation (3.1) other than the one we have found above. But, perhaps there are other solutions that differ from this one locally? The answer in the negative is a direct consequence of the argument presented in the local theorem of Picard.

3.6 Exercises

Exercise 3.1. What is the name of the famous mathematician, who was dumped by his sweetheart because in paeans on her beauty he never used the phrase 'one and only'?

Exercise 3.2. Check that the subset of $C[t_0-\alpha, t_0+\alpha]$, composed of functions having values in a closed ball, is closed.

Exercise 3.3. Show that there is no L satisfying (3.6). **Hint:** Think of $u = \frac{1}{n^2}$.

Exercise 3.4. Show that the function $f(u) = 1 + u^2$ is locally Lipschitz continuous but not globally Lipchitz continuous.

Exercise 3.5. Show that any function $f\colon [a,b] \to \mathbb{R}$ that is continuously differentiable is Lipchitz continuous. **Hint:** Use the mean-value theorem of Lagrange.

Exercise 3.6. Assume that $\|(x,y)\|_{\max} \le r$ and $\|(\tilde{x},\tilde{y})\|_{\max} \le r$ to show that

$$\begin{aligned}\|(ax - bxy, \, -cy + dxy) - (a\tilde{x} - b\tilde{x}\tilde{y}, \, -c\tilde{y} + d\tilde{x}\tilde{y})\|_{\max} \\ \le \max\{a + 2br, c + 2dr\}\|(x,y) - (\tilde{x},\tilde{y})\|_{\max}.\end{aligned}$$

Exercise 3.7. Find constants c and C having the following property: for all $x, y \in \mathbb{R}$ we have

$$c\max\{|x|,|y|\} \le \sqrt{x^2 + y^2} \le C\max\{|x|,|y|\}.$$

This implies that the Euclidean metric is equivalent to the metric related to the maximum norm in $\mathbb{R}^2$ (see the next chapter).

Exercise 3.8. Check that formula (3.8) defines a metric in the space of continuous functions.

Exercise 3.9. Prove that the right-hand side in (3.12) is finite and that d_ψ is a metric in $C[a,b]$.

Exercise 3.10. Arguing as in Section 3.2, prove completeness of the space from Section 1.5.

Exercise 3.11. Prove that d_ω introduced in the proof of the global theorem of Picard is well defined (in particular $d_\omega(x,y) < \infty$ for all $x, y \in C_\omega(\mathbb{R}^+)$) in $C_\omega(\mathbb{R}^+)$, and that it is a metric.

Exercise 3.12. Prove (3.20).

Exercise 3.13. Let $x\colon [a,b] \to \mathbb{R}^k$ be a continuous function. Show that

$$\left\| \int_a^b x(t)\, \mathrm{d}t \right\| \le \int_a^b \|x(t)\|\, \mathrm{d}t.$$

Exercise 3.14. Let y and ϕ be real continuous functions on $[0,1]$. Show that there is precisely one function $x \in C[0,1]$ (the space of real continuous functions on $[0,1]$) such that

$$x(t) - \int_0^t \phi(t-s)x(s)\, \mathrm{d}s = y(t), \qquad t \in [0,1].$$

Hint: Consider $C[0,1]$ with Bielecki distance

$$d_\lambda(x_1,x_2) = \sup_{t\in[0,1]} \mathrm{e}^{-\lambda t}|x_1(t) - x_2(t)|$$

and the map T in this space given by

$$(Tx)(t) = y(t) + \int_0^t \phi(t-s)x(s)\, \mathrm{d}s, \qquad t \in [0,1].$$

Exercise 3.15. Check that there is precisely one real continuous function x on $[0,1]$ such that

$$x(t) + \int_0^t [\sin(ts) + \cos^4(s+t)]x(s)\, \mathrm{d}s = \frac{1}{1+t^2}, \qquad t \in [0,1].$$

Repeat the exercise for the equation

$$x(t) - \int_0^t [\sin(t+s) + \cos^4(st)]x(s)\, \mathrm{d}s = \frac{t^2-1}{t^2-3t+2}, \qquad t \in [0,1),$$

and for the equation

$$x(t) + \sin t^2 = \tfrac{1}{2}x(t/3) + \tfrac{1}{3}x(t/2) + \mathrm{e}^{\pi t}, \qquad t \in [0,1].$$

Exercise 3.16. Consider the space of polynomials $P[0,1]$ treated as functions defined on $[0,1]$. In this space, given C and $\alpha \in \mathbb{R}$, consider the map T defined by

$$(Tx)(t) = C + \alpha \int_0^t x(s)\,\mathrm{d}s, \qquad t \in [0,1], x \in P[0,1].$$

Check to see that T maps $P[0,1]$ into itself. Show also that T satisfies (2.1) with $q := \frac{|\alpha|}{\lambda}$ provided that P is equipped with the Bielecki-type distance. Since λ may be chosen larger than $|\alpha|$, the Banach fixed point theorem tells us that there is a polynomial such that

$$x(t) = C + \alpha \int_0^t x(s)\,\mathrm{d}s, \qquad t \in [0,1].$$

But, wait a minute, such a polynomial would need to be equal to the exponential function, which is impossible! What does this tell you about the space of polynomials equipped with Bielecki-type norms? See Section 4.4.2.

Exercise 3.17. ▲ Let a be a real continuous function on $[0,1]$. Show that for every α, β and $\gamma \in \mathbb{R}$, there is precisely one twice continuously differentiable function on this interval such that

$$x''(t) + a(t)x'(t) = \gamma x(t), \qquad t \in [0,1] \qquad \text{and} \qquad x(0) = \alpha, x'(0) = \beta. \tag{3.24}$$

Hint. A solution to the above problem satisfies

$$x'(t) = \beta \mathrm{e}^{-\int_0^t a(s)\,\mathrm{d}s} + \gamma \int_0^t \mathrm{e}^{-\int_s^t a(u)\,\mathrm{d}u} x(s)\,\mathrm{d}s, \qquad t \in [0,1].$$

In the space of real continuous functions on $[0,1]$ consider the map given by

$$(Tx)(t) := \alpha + \beta \int_0^t \mathrm{e}^{-\int_0^s a(u)\,\mathrm{d}u}\,\mathrm{d}s + \gamma \int_0^t \int_0^s \mathrm{e}^{-\int_u^s a(v)\,\mathrm{d}v} x(u)\,\mathrm{d}u\,\mathrm{d}s,\ t \in [0,1].$$

Exercise 3.18. Let $(\alpha_n)_{n\geq 1}$ and $(\beta_n)_{n\geq 1}$ be two convergent sequences, and let $\alpha = \lim_{n\to\infty} \alpha_n$ and $\beta = \lim_{n\to\infty} \beta_n$. Prove that unique solutions of

$$x''(t) + a(t)x'(t) = \gamma x(t), \qquad t \in [0,1] \qquad \text{and} \qquad x(0) = \alpha_n, x'(0) = \beta_n$$

converge to those of (3.24). This is a particular case of a theorem on continuous dependence of solutions of differential equations on initial data. **Hint.** Use Exercise 2.6.

☞ CHAPTER SUMMARY

We prove two versions of the celebrated theorem of Picard: the local and the global existence and uniqueness results for differential equations. We use

Banach's principle as a main tool in our analysis, and this makes us realize that it is the completeness of the space of continuous functions that is the reason for the existence of solutions to differential equations. In the mean-time we get acquainted with the notion of equivalent metrics and learn that, in proving the existence of a fixed point of a map, it is sometimes more convenient to use one norm and sometimes another, equivalent one.

4
Banach Spaces

4.1 Definition of Banach Space

A Banach space, the central object of functional analysis, is by definition a linear, normed and complete space. In this chapter, we provide a step-by-step introduction and explanation of the notions just mentioned.

4.1.1 Linear Space

Let $\mathbb{X}$ be a set; its elements will be denoted x, y, z, etcetera, and called vectors. The triple $(\mathbb{X}, +, \cdot)$, where $+$ is a map $+ \colon \mathbb{X} \times \mathbb{X} \to \mathbb{X}, (x, y) \mapsto x + y$ and $\cdot$ is a map $\cdot \colon \mathbb{R} \times \mathbb{X} \to \mathbb{X}, (\alpha, x) \mapsto \alpha x$, is said to be a (real) linear space (or a vector space) if the following conditions are satisfied:[1,2]

(a1) for all $x, y, z \in \mathbb{X}$ we have $(x + y) + z = x + (y + z)$,
(a2) there is a $\Theta \in \mathbb{X}$ such that $x + \Theta = x$ for all $x \in \mathbb{X}$,
(a3) for any $x \in \mathbb{X}$ there exists an $x' \in \mathbb{X}$ such that $x + x' = \Theta$,
(a4) $x + y = y + x$ for all $x, y \in \mathbb{X}$,
(m1) $\alpha(\beta x) = (\alpha\beta)x$ for all $\alpha, \beta \in \mathbb{R}, x \in \mathbb{X}$,
(m2) $1x = x$ for all $x \in \mathbb{X}$,
(d) $\alpha(x + y) = \alpha x + \alpha y$ and $(\alpha + \beta)x = \alpha x + \beta x$ for all scalars $\alpha, \beta \in \mathbb{R}$ and vectors $x, y \in \mathbb{X}$.

The element $\Theta \in \mathbb{X}$ of point (a2) is unique, is called the zero vector, or simply zero, and often denoted simply 0 (although the zero vector should be distinguished from the number zero); it turns out also that for each $x \in \mathbb{X}$, $0x$

[1] Although, strictly speaking, a vector space is an ordered triple, more often than not, if this does not lead to misunderstandings, we will say that $\mathbb{X}$ alone is a linear space.
[2] The list groups linked conditions together. (a1)–(a4) describe (commutative) group structure, (m1) and (m2) describe multiplication by scalars, and (d) is the distributive property.

(where 0 is the zero number) is the zero vector. Also x' of point (a3) is said to be opposite to x, and it can be checked that $x' = (-1)x$. Hence, it is customary to denote this vector by $-x$ rather than by x' (and in what follows we adhere to this practical custom).

If you want a good example of a linear space, think of the space of real functions on a given set S. Addition of functions and multiplication by scalars is provided here by the standard formulae:

$$(x + y)(p) = x(p) + y(p), \qquad (\alpha x)(p) = \alpha x(p),$$

where x and y are functions and $p \in S$. The space of $n \times m$ matrices is a special case of this example: any $n \times m$ matrix can be thought of as a function defined on the set S equal to the Cartesian product $\{1, \dots, n\} \times \{1, \dots, m\}$. Similarly, the space of sequences is just the space of functions defined on the set of integers.

In the space of functions, the zero vector is the function that equals zero at all $p \in S$, and the vector x' that is opposite to the given function x is defined by $x'(p) = -x(p), p \in S$.

A subspace (in the algebraic sense) of a linear space $\mathbb{X}$ is its subset $\mathbb{Y}$ with the following property: conditions $x, y \in \mathbb{Y}, \alpha \in \mathbb{R}$ imply $x + y \in \mathbb{Y}$ and $\alpha x \in \mathbb{Y}$. There are many good examples of subspaces of the space of functions on a given set: for instance, the subspace of bounded functions and the subspace of functions that vanish on a given, fixed point $p_0 \in S$. (To recall, a function defined on S is said to be bounded if there is a constant M such that $|f(p)| \leq M$ for all $p \in S$.) All we are saying here is that a sum of two bounded functions is bounded, and that by multiplying a bounded function by a scalar we obtain another bounded function. Similarly, if two functions vanish at a point, then so do their sum and all their products by a scalar.

Likewise, one concludes that if S is a topological space then the space of continuous functions is a subspace of all functions on S, and if S is a measurable space then so is the space of measurable functions. For a yet another example, consider the case where S is the set of reals: then the subspace of even (odd) functions is a subspace of all functions on $\mathbb{R}$.

It is clear from the definition that a subspace of a linear space is a linear space itself with addition and multiplication inherited from the larger space. A final terminological remark: expressions of the form $\alpha x + \beta y$ are conveniently referred to as linear combinations of x and y.

4.1.2 Normed Spaces

A vector space $\mathbb{X}$ is said to be normed if there is a function

$$\| \cdot \| : \mathbb{X} \to \mathbb{R},$$

termed a norm, and interpreted as the vector's length, enjoying the following properties. For all $x, y \in \mathbb{X}$ and $\alpha \in \mathbb{R}$,

(n1) $\|x\| \geq 0$,
(n2) $\|x\| = 0$ if and only if x is the zero vector,
(n3) $\|\alpha x\| = |\alpha| \, \|x\|$,
(n4) $\|x + y\| \leq \|x\| + \|y\|$.

As it was in the case of a metric, the fourth condition here is termed the triangle inequality; the other three are also self-explanatory if we remember that $\|x\|$ is to represent the length of $\|x\|$.

Readers can check their basic mathematical skills by proving (Exercise 4.1) that conditions (n3–n4) imply that for all $x, y \in \mathbb{X}$, we have

$$\big| \|x\| - \|y\| \big| \leq \|x \pm y\|. \tag{4.1}$$

It should be noted here that if $\|\cdot\|$ is a norm then the formula $d(x, y) = \|x - y\|$ defines a metric (see Exercise 4.2). In other words, each normed space is a metric space. Hence, in a linear space it is meaningful to speak of topological notions such as convergence, open and closed sets, etcetera. But the structure of a linear space is much richer than the structure of a metric space: to state the most obvious reason, the normed space is also equipped with the linear structure, and the metric in it is connected with this structure in a natural yet intricate way. It is because of this richness that there are so many things 'going on' in normed spaces, and even more so in Banach spaces.

4.2 Examples of Normed Spaces

Let's have a closer look at the classical normed spaces.

4.2.1 The Space c of Convergent Sequences

This space is composed of numerical sequences $x = (\xi_i)_{i \geq 1}$ that converge to a finite limit at infinity, that is, for a member of c there exists $\lim_{i \to \infty} \xi_i$. It is amply clear that this is a linear space: if sequences $x = (\xi_i)_{i \geq 1}$ and $y = (\eta_i)_{i \geq 1}$ have limits, then for all scalars α and β so does the sequence $\alpha x + \beta y$. Each element $x \in c$ is also bounded: if g is its limit, then all but a finite number of its elements lie in the interval $(g - 1, g + 1)$. Hence,

$$\sup_{i \geq 1} |\xi_i| \leq \max\{M, |g| + 1\} < \infty,$$

where M is the maximal absolute value of elements that lie outside of $(g-1, g+1)$. Therefore, to an $x \in c$ we can assign

$$\|x\| = \sup_{i \geq 1} |\xi_i| < \infty,$$

and it is easy to check that $\| \cdot \|$ is a norm.

4.2.2 The Space c_0 of Sequences Converging to 0

If we have two sequences, say $x = (\xi_i)_{i \geq 1}$ and $y = (\eta_i)_{i \geq 1}$, that converge to 0, then – for all scalars α and β – the sequence $\alpha x + \beta y$ converges to 0 as well. This shows that, in the algebraic sense, the set of sequences that converge to 0 is a subspace of c. This subspace is customarily denoted c_0. With the norm inherited from c, c_0 is a normed space.

Moreover, c_0 is a closed subset of c. To see this, assume that $x_n \in c_0$ converges to an $x = (\xi_i)_{i \geq 1} \in c$ and let $\epsilon > 0$ be given. Moreover, let N be a natural number so big that $\|x_n - x\| < \frac{\epsilon}{2}$ for $n \geq N$. Since $x_N = \left(\xi_{N,i}\right)_{i \geq 1}$, by assumption, converges to 0, we can choose an m so that $|\xi_{N,i}| < \frac{\epsilon}{2}$ for all $i \geq m$. On the other hand, for all i, we have $|\xi_{N,i} - \xi_i| < \frac{\epsilon}{2}$. It follows that $|\xi_i| < \epsilon$ for all $i \geq m$. In other words, all but a finite number of elements of $(\xi_i)_{i \geq 1}$ are smaller than ϵ. Since ϵ is arbitrary here, we have shown that $(\xi_i)_{i \geq 1}$ converges to 0.

Therefore, c_0 is a subspace of c in a stronger sense: it is not only 'closed' in the algebraic sense, but also in the topological sense. Previously we have established that taking products and sums does not lead out of this space; now we know also that taking a limit does not lead us out of this space. This information will be important later on.

4.2.3 The Space $C[a, b]$ of Continuous Functions on $[a, b]$

As we have mentioned above, the set of all real continuous functions that are continuous on a closed interval $[a, b]$ (where, of course, $a < b$) is a subspace of all functions defined there. This space will be denoted $C[a, b]$ (this should not lead to misunderstandings – we are dealing with a particular case of the space of Section 3.2). From Section 3.2 we know that $C[0, 1]$ is a metric space. As it turns out, the metric in $C[0, 1]$ comes from a norm, termed the supremum norm. To explain, we recall a theorem of basic calculus/real analysis saying that each continuous function defined on a compact set (and a closed interval is a compact set) attains a maximum and a minimum, each at least once. Therefore, for $x \in C[a,b]$ there is a $t_0 \in [a,b]$ such that

$$\max_{t \in [a,b]} |x(t)| = |x(t_0)|.$$

It is easy to see that

$$x \mapsto \|x\| := \max_{t \in [a,b]} |x(t)|$$

is a norm. Moreover, this norm is related to the supremum metric d as follows:

$$d(x,y) = \|x - y\|.$$

This means that $C[a,b]$ is a normed vector space. Also, as we know from Section 3.2, this space is complete. Hence, we have a first example of a Banach space – see Section 4.3.

4.2.4 The Space ℓ^1 of Absolutely Summable Sequences

We say that a sequence $(\xi_i)_{i \geq 1}$ of real numbers is absolutely summable if the sum

$$\|(\xi_i)_{i \geq 1}\| := \sum_{i=1}^{\infty} |\xi_i|$$

is finite. It is not accidental that we have used the notation $\| \cdot \|$ here – $\| \cdot \|$ is indeed a norm in the space ℓ^1 of absolutely summable sequences. Of course, it is fitting to check first that ℓ^1 is a linear space – this, however, cannot be doubted, for a linear combination of two absolutely summable sequences is also absolutely summable. The reader will also not encounter serious difficulties in checking that $x \mapsto \|x\|$ as defined above has all the properties of a norm.

4.2.5 The Space $L^1(\mathbb{R})$ of Lebesgue Integrable Functions

The space $L^1(\mathbb{R})$ is a continuous counterpart of ℓ^1: it is the space of absolutely integrable functions on $\mathbb{R}$ with the norm

$$\|x\| = \int_{\mathbb{R}} |x|. \tag{4.2}$$

Strictly speaking, $L^1(\mathbb{R})$ is not composed of functions but of equivalence classes of functions. The point is that condition $\|x\| = 0$ *does not* imply $x \equiv 0$, but merely that x equals 0 almost everywhere (that is, everywhere except on a set of measure zero). If we agree that two functions are in relation if they differ on a set of measure zero, we obtain an equivalence relation. (For example, if $x(t)$ is zero for all irrational t and 1 for all rational t, then x is in relation with the function that equals zero everywhere.) Moreover, the value of $\int_{\mathbb{R}} |x|$ is the same for all functions that are in this equivalence relation. Hence, $x \mapsto \int_{\mathbb{R}} |x|$ can be thought of as a function that is defined on equivalence classes, and is a norm in the space of these equivalence classes (the sum of two classes is the class corresponding to any sum of functions from the original classes; likewise for multiplication by scalars). In particular, the zero vector in $L^1(\mathbb{R})$ is the equivalence class of functions that are in relation with the function identically equal zero, that is, the class of functions that vanish almost everywhere.

All the properties required for a norm are almost immediate in the case of $\| \cdot \|$ defined in (4.2). For example, to check the triangle inequality all one needs to know is the triangle inequality for the norm in $\mathbb{R}$ and that $\int (x + y) = \int x + \int y$ (i.e., that the integral is a linear map).

In what follows, we will adhere to the apparently common custom and refer to $L^1(\mathbb{R})$ as the space of absolutely integrable functions, and not as the space of equivalence classes. Usually this does not lead to misunderstandings but the distinctions should be kept in the back of our minds.

4.3 Examples of Banach Spaces

4.3.1 The Space c of Convergent Sequences

The space c of Section 4.2.1 is complete. For, suppose $(x_n)_{n\geq 1}$ is a Cauchy sequence in c and let $x_n = \left(\xi_{n,i}\right)_{i\geq 1}, n \geq 1$. For any $\epsilon > 0$ there is thus an $N \in \mathbb{N}$ such that

$$\sup_{i\geq 1} |\xi_{n,i} - \xi_{m,i}| = \|x_n - x_m\| < \epsilon$$

for $n, m \geq N$. Fix, for the time being, an $i \in \mathbb{N}$. The obvious inequality

$$|\xi_{n,i} - \xi_{m,i}| \leq \|x_n - x_m\| \tag{4.3}$$

convinces us, in the light of what was already said, that the numerical sequence $\left(\xi_{n,i}\right)_{n\geq 1}$ satisfies the Cauchy condition, and thus has a limit (because $\mathbb{R}$ is a complete space). Let ξ_i denote the limit of $\left(\xi_{n,i}\right)_{n\geq 1}$. Clearly, the sequence $x = (\xi_i)_{i\geq 1}$ composed of limits ξ_i obtained in the procedure described above is a natural candidate for the limit of $(x_n)_{n\geq 1}$.

We are left with showing that $x \in c$ and that $\lim_{n\to\infty} x_n = x$. We know that for any $\epsilon > 0$ an N can be chosen in such a way that

$$|\xi_{n,i} - \xi_{m,i}| < \frac{\epsilon}{2}, \quad \text{for all } i \in \mathbb{N}, \tag{4.4}$$

as long as $n, m \geq N$. Letting $i \to \infty$ we see that

$$|\xi_n - \xi_m| \leq \frac{\epsilon}{2} < \epsilon, \qquad \text{for } n, m \geq N,$$

showing that $(\xi_i)_{i\geq 1}$ satisfies the Cauchy condition, and thus is convergent, that is, it belongs to c. On the other hand, letting $m \to \infty$ in (4.4) (for each i separately), we obtain

$$|\xi_{n,i} - \xi_i| \leq \frac{\epsilon}{2}, \quad \text{for all } i \in \mathbb{N},$$

and this shows that

$$\|x_n - x\| \leq \frac{\epsilon}{2} < \epsilon.$$

This, however, proves that x is the limit of $(x_n)_{n\geq 1}$.

4.3.2 The Space c_0 of Sequences Converging to Zero

As we have seen, c_0 of Section 4.2.2 is a subspace of c in the algebraic sense and that it inherits a norm from c. We have also proved that c_0 is a closed subset of c. Theorem 2.1 tells thus that c_0 itself is complete, and this means that c_0 is a Banach space.

4.3.3 The Space $C[a,b]$ of Continuous Functions on $[a,b]$

In Section 3.2 we learned that $C[a,b]$, when equipped with the supremum distance is a complete space, and in Section 4.2.3 we found out that it is also a linear normed space. Furthermore, the supremum distance is inherited from the natural supremum norm in $C[a,b]$. All these facts together show that $C[a,b]$ is a Banach space.

4.3.4 The Space ℓ^1 of Absolutely Summable Sequences

It has already been established in Section 4.2.4 that ℓ^1 is a normed vector space. To show that this space is complete, let's assume that $(x_n)_{n\geq 1}$ is a fundamental sequence in c and let $x_n = \left(\xi_{n,i}\right)_{i\geq 1}$. For any $\epsilon > 0$ one can find an N such that

$$\sum_{i=1}^{\infty} |\xi_{n,i} - \xi_{m,i}| < \epsilon,$$

as long as $n, m \geq N$. Similarly as in the case of c, we use the inequality[3]

$$|\xi_{n,i} - \xi_{m,i}| \leq \|x_n - x_m\| \tag{4.5}$$

to check that for any i, the numerical sequence $(\xi_{n,i})_{n \geq 1}$ converges because it satisfies the Cauchy condition in a complete space.

And, once again, $x = (\xi_i)_{i \geq 1}$, where $\xi_i = \lim_{n \to \infty} \xi_{i,n}$, is a natural candidate for the limit of $(x_n)_{n \geq 1}$. By assumption, we know that for any $\epsilon > 0$, one can choose a natural $N = N(\epsilon)$ so that

$$\sum_{i=1}^{\infty} |\xi_{n,i} - \xi_{m,i}| < \frac{\epsilon}{2},$$

as long as $n, m \geq N$. This shows that for any natural k we have $\sum_{i=1}^{k} |\xi_{n,i} - \xi_{m,i}| < \frac{\epsilon}{2}$. Letting $m \to \infty$, we conclude that, for all k, $\sum_{i=1}^{k} |\xi_{n,i} - \xi_i| \leq \frac{\epsilon}{2}$, and so

$$\sum_{i=1}^{\infty} |\xi_{n,i} - \xi_i| \leq \frac{\epsilon}{2} < \epsilon. \tag{4.6}$$

For $\epsilon = 1$ and $N = N(1)$, this yields

$$\sum_{i=1}^{\infty} |\xi_i| \leq \sum_{i=1}^{\infty} \left(|\xi_i - \xi_{N,i}| + |\xi_{N,i}|\right) < 1 + \|x_N\| < \infty.$$

This, however, shows that $x \in \ell^1$. The inequality (4.6) establishes now that $\|x_n - x\| < \epsilon$ for all sufficiently large n. And this means simply that x is a limit of $(x_n)_{n \geq 1}$.

4.3.5 The Space $L^1(\mathbb{R})$ of Absolutely Integrable Functions

The proof of the fact that $L^1(\mathbb{R})$ is a Banach space when equipped with the norm

$$\|x\| := \int_{\mathbb{R}} |x(t)| \, dt$$

differs from that of the completeness of ℓ^1. The main reason for this is that in $L^1(\mathbb{R})$ there is no counterpart of (4.3) or (4.5). In fact, convergence in $L^1(\mathbb{R})$

[3] Although this inequality is apparently identical to that of Section 4.3.1, they are in fact quite different. In other words, on the right-hand sides of these inequalities we have two different norms, and thus the arguments used to establish these inequalities are also different.

does not imply pointwise convergence (cf. Exercise 11.6). Hence, we cannot follow the examples presented above and, given a Cauchy sequence $(x_n)_{n\geq 1}$ of elements of $L^1(\mathbb{R})$, define the searched for x as $x(t) := \lim_{n\to\infty} x_n(t), t \in \mathbb{R}$. The latter limit can exist almost nowhere!

Therefore, we take a different route, and use one of the basic results of the theory of integration, called the monotone convergence theorem. This theorem is devoted to the situation where we are given a sequence $(y_n)_{n\geq 1}$ of real measurable functions defined on $\mathbb{R}$ such that $0 \leq y_n(t) \leq y_{n+1}(t)$ for all $t \in \mathbb{R}$ and $n \geq 1$. Then, for each $t \in \mathbb{R}$, the numerical sequence $(y_n(t))_{n\geq 1}$ is non-decreasing and either has a finite limit (if and only if it is bounded) or converges to infinity. Therefore, one may think of y, defined as $y(t) := \lim_{n\to\infty} y_n(t)$, as a function with values in $\mathbb{R}^+ \cup \{\infty\}$. A similar argument shows that the limit of the numerical sequence $\left(\int_{\mathbb{R}} y_n(t)\,\mathrm{d}t\right)_{n\geq 1}$ exists (but may be infinite). The monotone convergence theorem connects this limit with the integral $\int_{\mathbb{R}} y(t)\,\mathrm{d}t$ by saying that

$$\lim_{n\to\infty} \int_{\mathbb{R}} y_n(t)\,\mathrm{d}t = \int_{\mathbb{R}} y(t)\,\mathrm{d}t$$

(note that both sides of this equality may be infinite).

So prepared, we consider a Cauchy sequence $(x_n)_{n\geq 1}$ of elements of $L^1(\mathbb{R})$, and recursively define its subsequence $(x_{n_i})_{i\geq 1}$ as follows. Its first element, x_{n_1}, is defined by requiring that n_1 is an index such that $\|x_n - x_m\| < 1$ for all $n, m \geq n_1$. Next, if $x_{n_1}, \ldots, x_{n_k}$ are already defined, we pick n_{k+1} larger than $n_1, \ldots, n_k$ and such that $\|x_n - x_m\| < \frac{1}{2^k}$ for all $n, m \geq n_{k+1}$. This completes our construction of $(x_{n_i})_{i\geq 1}$.

Then, we think of

$$y_n(t) := |x_{n_1}(t)| + \sum_{k=2}^{n} |x_{n_k}(t) - x_{n_{k-1}}(t)|, \qquad t \in \mathbb{R}, n \geq 1.$$

Clearly, $(y_n)_{n\geq 1}$ is a non-decreasing sequence of measurable functions, as in the monotone convergence theorem. It follows that for y defined as

$$y(t) := \lim_{n\to\infty} y_n(t), \qquad t \in \mathbb{R},$$

we have

$$\begin{aligned}\int_{\mathbb{R}} y(t)\,\mathrm{d}t &= \int_{\mathbb{R}} |x_{n_1}(t)|\,\mathrm{d}t + \lim_{n\to\infty} \sum_{k=2}^{n} \int_{\mathbb{R}} |x_{n_k}(t) - x_{n_{k-1}}(t)|\,\mathrm{d}t \\ &= \|x_{n_1}\| + \lim_{n\to\infty} \sum_{k=2}^{n} \|x_{n_k} - x_{n_{k-1}}\|,\end{aligned}$$

and by the way $(x_{n_i})_{i\geq 1}$ was constructed we see that

$$\lim_{n\to\infty}\sum_{k=2}^{n}\|x_{n_k}-x_{n_{k-1}}\|\leq \lim_{n\to\infty}\sum_{k=2}^{n}\frac{1}{2^{k-2}}=2.$$

(This is because $n_k\geq n_{k-1}$, which implies $\|x_{n_k}-x_{n_{k-1}}\|\leq\frac{1}{2^{k-2}}$ by definition of n_{k-1}.) Consequently,

$$\int_{\mathbb{R}} y(t)\,\mathrm{d}t<\infty,$$

and thus the measure of t for which $y(t)=\infty$ has to be zero. In other words, for all but t in a set of measure zero, the series

$$|x_{n_1}(t)|+\sum_{k=2}^{\infty}|x_{n_k}(t)-x_{n_{k-1}}(t)|$$

converges (to a finite sum). Therefore, the series

$$x_{n_1}(t)+\sum_{k=2}^{\infty}(x_{n_k}(t)-x_{n_{k-1}}(t)) \tag{4.7}$$

converges also, for all t except for those belonging to the set of measure zero. Let $x(t)$ denote its sum for such t and let $x(t)=0$ for the remaining t. Cleary, $|x(t)|\leq y(t)$ and thus $\int_{\mathbb{R}}|x(t)|\,\mathrm{d}t<\infty$, that is, x belongs to $L^1(\mathbb{R})$. Since the ℓth partial sum of the series (4.7) equals $x_{n_\ell}, \ell\geq 1$, we see that $\lim_{\ell\to\infty}x_{n_\ell}=x(t)$ for all t save those in the set of measure zero.

To complete the proof we need another key result of the theory of integration, that is, the dominated convergence theorem. This theorem may be rephrased as follows. Assume that $(z_n)_{n\geq 1}$ is a sequence of elements of $L^1(\mathbb{R})$ and that there is a member, say $\tilde{z}$, of this space, such that $|z_n(t)|\leq\tilde{z}(t)$ for all $n\geq 0$ and $t\in\mathbb{R}$. Assume also that $z(t)=\lim_{n\to\infty}z_n(t)$ exists save for t in a set of measure zero. Then, upon extending the definition of $z(t)$ to be zero on the set where the limit does not exist, we obtain

$$\lim_{n\to\infty}\int_{\mathbb{R}}|z_n(t)-z(t)|\,\mathrm{d}t=0.$$

In particular, by $\left|\int_{\mathbb{R}}z_n(t)\,\mathrm{d}t-\int_{\mathbb{R}}z(t)\,\mathrm{d}t\right|\leq\int_{\mathbb{R}}|z_n(t)-z(t)|\,\mathrm{d}t$, we have

$$\lim_{n\to\infty}\int_{\mathbb{R}}z_n(t)\,\mathrm{d}t=\int_{\mathbb{R}}z(t)\,\mathrm{d}t.$$

Coming back to our proof, we are now ready to show that $(x_n)_{n\geq 1}$ converges to x. Indeed, given $\epsilon>0$ we may find an n_0 such that, for $n,m\geq n_0$, $\|x_n-x_m\|<4/5\epsilon$. Since $\lim_{i\to\infty}n_i=\infty$ (as, n_{k+1} is larger than $n_1,\ldots,n_k$), we have $n_i \geq n_0$ for all but a finite number of i. For such i and any fixed $n \geq n_0$,

$$\int_{\mathbb{R}} |x_n(t) - x_{n_i}(t)|\,\mathrm{d}t < \frac{4}{5}\epsilon. \tag{4.8}$$

Letting $z_i(t) := |x_n(t) - x_{n_i}(t)|$ we have $\lim_{i\to\infty} z_i(t) = |x_n(t) - x(t)|$ for t outside of a set of measure zero. Moreover, for any t,

$$|z_i(t)| \leq |x_n(t)| + |x_{n_i}(t)|,$$

whereas

$$\begin{aligned} |x_{n_i}(t)| &= |x_{n_1}(t) + (x_{n_2} - x_{n_1})(t) + \cdots + (x_{n_i} - x_{n_{i-1}})(t)| \\ &\leq |x_{n_1}|(t) + |x_{n_2} - x_{n_1}|(t) + \cdots + |x_{n_i} - x_{n_{i-1}}|(t) \leq y(t). \end{aligned}$$

Hence,

$$|z_i(t)| \leq |x_n(t)| + y(t),$$

and we know that $\int_{\mathbb{R}}[|x_n(t)| + y(t)]\,\mathrm{d}t < \infty$. Thus, by the dominated convergence theorem, the left hand side in (4.8) converges, as $i \to \infty$, to

$$\int_{\mathbb{R}} |x_n(t) - x(t)|\,\mathrm{d}t,$$

showing that for $n \geq n_0$, $\|x_n - x\| \leq \frac{4}{5}\epsilon < \epsilon$. Since $\epsilon > 0$ is arbitrary, the proof is complete.

4.4 Normed Spaces that Are Not Complete

4.4.1 $C[0,2]$ with 'Improper' Norm

Let's turn to an example of a normed space that is not complete. To this end consider $C[0,2]$ with the norm

$$\|x\| = \int_0^2 |x(s)|\,\mathrm{d}s.$$

We know that $C[0,2]$ is a vector space and we know that the so-defined $\|\cdot\|$ is a norm. Our goal is to show that $C[0,2]$ with this norm is not complete. For this, we need to find a fundamental sequence in $C[0,2]$ that does not converge. Our candidate sequence is defined as follows: $x_n(s) = \min\{s^n, 1\}, s \in [0,2]$; see Figure 4.1.

Since

$$\|x_n - x_m\| = \int_0^1 (s^n - s^m)\,\mathrm{d}s = \frac{1}{n+1} - \frac{1}{m+1},$$

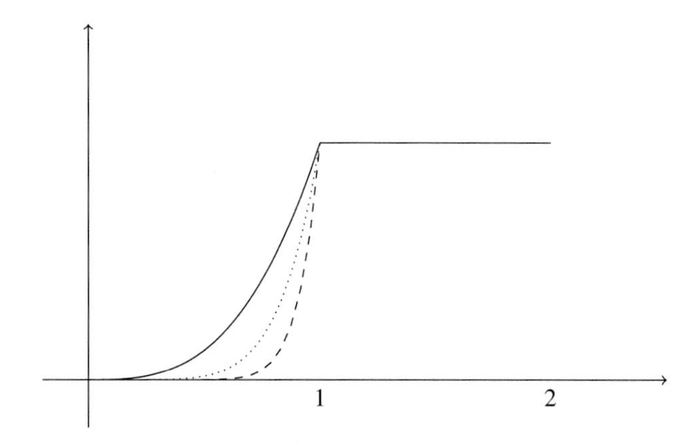

Figure 4.1 Graphs of x_n for $n = 3$ (solid line), $n = 6$ (dotted) and $n = 11$ (dashed); in the interval $[1,2]$ these graphs coincide.

as long as $m \geq n$, it is clear that the Cauchy condition is satisfied. However, $(x_n)_{n\geq 1}$ does not converge. To see this, let's think of how its limit, say x, would need to look. For $s > 1$, this x would need to equal 1, for otherwise we would have $\|x_n - x\| \geq \int_1^2 |1 - x(s)|\,\mathrm{d}s > 0$ for each $n \geq 1$. Likewise, we would need to have $x(s) = 0$ for $s < 1$, for otherwise for some $\delta \in (0,1)$ we would have $\lim_{n\to\infty} \|x_n - x\| \geq \lim_{n\to\infty} \int_0^\delta |x(s) - s^n|\,\mathrm{d}s = \int_0^\delta |x(s)|\,\mathrm{d}s > 0$ (since s^n converges uniformly, as $n \to \infty$, to 0 on any interval $[0,\delta]$, $\delta \in (0,1)$). There is, however, no continuous function x that simultaneously satisfies $x(s) = 0, s < 1$ and $x(s) = 1, s > 1$.

4.4.2 Space of Polynomials

Here is another example of a normed space that is not complete. Let $\mathbb{X}$ be the space of polynomials, defined on the interval $[0,1]$, that is, of functions of the form

$$x(t) = a_0 + a_1 t + \cdots + a_n t^n,$$

where n is a natural number, which varies with x, and $a_i, i = 0, \ldots, n$ are real numbers. It is easy to check that $\mathbb{X}$ is a vector space. The map

$$x \mapsto \|x\| := \sup_{t\in[0,1]} |x(t)|$$

is a norm for $\mathbb{X}$. As it turns out, however, this space is not complete.

To see this, let a sequence $(x_n)_{n\ge 1}$ of functions on $[0,1]$ be given by

$$x_n(t) = 1 + t + \frac{t^2}{2} + \cdots + \frac{t^n}{n!}, \qquad t \in [0,1], n \ge 1.$$

It is clear that all x_n are polynomials, that is, elements of $\mathbb{X}$. By (4.9), for all integers n and m, we also have

$$\begin{aligned}\|x_m - x_n\| &\le \sup_{t\in[0,1]} \int_0^t \frac{(t-s)^n}{n!} \mathrm{e}^s \, \mathrm{d}s + \sup_{t\in[0,1]} \int_0^t \frac{(t-s)^m}{m!} \mathrm{e}^s \, \mathrm{d}s \\ &\le \mathrm{e}^1 \int_0^1 \frac{s^n}{n!} \, \mathrm{d}s + \mathrm{e}^1 \int_0^1 \frac{s^m}{m!} \, \mathrm{d}s \le \mathrm{e}^1 \left(\frac{1}{(n+1)!} + \frac{1}{(m+1)!} \right).\end{aligned}$$

It is immediate from these inequalities that $(x_n)_{n\ge 1}$ is a Cauchy sequence (in the norm of $\mathbb{X}$, of course).

How can we convince ourselves that there is no limit of $(x_n)_{n\ge 1}$? The reason, and an obvious one for those who remember the definition of the exponential function in terms of a power series, is the fact that the function $x(t) = \mathrm{e}^t$ is the pointwise, and even uniform, limit of x_n (see Exercise 4.20). Hence, if there is a limit of x_n in the norm prescribed, it must coincide with this function, and we are left with showing that this function is not a member of $\mathbb{X}$. In other words, we need to argue that x is not a polynomial. This, however, is quite easy to see if we recall that the derivative of x is again x, and so derivatives of all orders of x are x also, and in particular none of them is identically zero. Polynomials, on the other hand, have the property that all but a finite number of their derivatives are zero.

4.5 Exercises

Exercise 4.1. Prove (4.1).

Exercise 4.2. Check to see that a normed space is automatically a metric space.

Exercise 4.3. ([1, p. 456]) Suppose vectors x, y in a normed space are such that $\|x + y\| = \|x\| + \|y\|$. Show that for all $\alpha, \beta \ge 0$, $\|\alpha x + \beta y\| = \alpha\|x\| + \beta\|y\|$.
Hint: Assuming $\alpha \ge \beta$, write $\alpha x + \beta y$ as $\alpha(x + y) - (\alpha - \beta)y$.

Exercise 4.4. Check that all the norms introduced in Section 4.2 are indeed norms.

Exercise 4.5. Let ℓ^∞ be the space of bounded sequences $(\xi_i)_{i\ge 1}$. Check to see that this is a normed linear space with norm $\|(\xi_i)_{i\ge 1}\| := \sup_{i\ge 1} |\xi_i|$.

Exercise 4.6. Let $\alpha \in (0,1)$ be given. Find the distance between sequences $(\alpha^i)_{i\ge 1}$ and $(\alpha^{i-1})_{i\ge 1}$ in c_0 and in ℓ^1. Check to see that as $\alpha \to 1$, both sequences 'escape' from c_0 and ℓ^1, but whereas in the former space the distance between them converges to zero, in the latter space the distance remains equal to 1. **Hint:** In the former space the distance is $1 - \alpha$; in the latter it is 1.

Exercise 4.7. A subset $\mathbb{Y}$ of a normed space $\mathbb{X}$ is said to be a (closed) subspace of $\mathbb{X}$ if and only if (a) linear combinations of members of $\mathbb{Y}$ belong to $\mathbb{Y}$ (i.e., $x, y \in \mathbb{Y}$ implies $\alpha x + \beta y \in \mathbb{Y}$ for any scalars α, β) and (b) for any sequence $(y_n)_{n\ge 1}$ of members of $\mathbb{Y}$, existence of the limit $y := \lim_{n\to\infty} y_n$ implies $y \in \mathbb{Y}$. Check to see that c_0 and c can be seen as subspaces of ℓ^∞.

Exercise 4.8. Given a subspace $\mathbb{Y}$ of a normed space $\mathbb{X}$, one constructs the related quotient space $\mathbb{X}/\mathbb{Y}$ as the set of equivalence classes of the equivalence relation $x \sim \widetilde{x}$ if and only if $x - \widetilde{x} \in \mathbb{Y}$; an equivalence class of $x \in \mathbb{X}$ is denoted $[x]$. Check to see that:

(a) by introducing addition and multiplication by formulae $[x] + [y] := [x + y]$, $\alpha[x] := [\alpha x]$, we make $\mathbb{X}/\mathbb{Y}$ a linear space,

(b) the function $[x] \mapsto |||[x]||| := \inf_{\widetilde{x}\in[x]} \|\widetilde{x}\|$ is a norm in $\mathbb{X}/\mathbb{Y}$ (termed a *quotient norm*).

Exercise 4.9. ▲ Check to see that the quotient norm of the class of a bounded sequence $(\xi_i)_{i\ge 1}$ in ℓ^∞/c_0 equals $\limsup_{i\to\infty} |\xi_i|$. **Hint:** To check that the norm does not exceed the lim sup, recall that for any $\epsilon > 0$ all but a finite number of $|\xi_i|$'s lie below $\limsup + \epsilon$. To prove the reverse inequality, recall that the lim sup is the same for all members of an equivalence class and that for any $\epsilon > 0$ infinitely many elements of $(|\xi|)_{i\ge 1}$ are larger than $\limsup_{i\to\infty} |\xi_i| - \epsilon$.

Exercise 4.10. Let $\mathbb{X}$ and $\mathbb{Y}$ be Banach spaces, and let $\mathbb{X} \times \mathbb{Y}$ be their Cartesian product, that is, the space of ordered pairs (x, y) where $x \in \mathbb{X}$ and $y \in \mathbb{Y}$. Check to see that, when equipped with the norm $\|(x, y)\| = \|x\| + \|y\|$, the Cartesian product is a Banach space also. Is the thesis the same if the norm is redefined to be $\sqrt{\|x\|^2 + \|y\|^2}$? What about the case in which $\|(x, y)\| = \max\{\|x\|, \|y\|\}$?

Exercise 4.11. Prove that ℓ^∞ with the norm defined above is a Banach space.

Exercise 4.12. Given $r > 0$, let l^1_r denote the space of sequences $(\xi_i)_{i\ge 1}$ such that

$$\|x\|_r := \sum_{i\geq 1} |\xi_i| r^i < \infty.$$

Prove that ℓ^1_r is a linear space, that the so-defined $\|\cdot\|_r$ is a norm, and that ℓ^1_r with this norm is a Banach space.

Exercise 4.13. Let E be the space of $(\xi_i)_{i\geq 1}$ such that $\sup_{n\geq 1} \frac{\sqrt{\xi_1^2+\cdots+\xi_n^2}}{n} < \infty$. Show that, when equipped with the norm

$$\|(\xi_i)_{i\geq 1}\| = \sup_{n\geq 1} \frac{\sqrt{\xi_1^2+\cdots+\xi_n^2}}{n},$$

E is a Banach space.

Exercise 4.14. Let $\mathbb{X}$ be the space of continuous functions $x\colon [0,2] \to \mathbb{R}$ such that $x(1) = \frac{1}{2}(x(0) + x(2))$. Show that when equipped with $\|x\| = \max_{t\in[0,2]} |x(t)|$, $\mathbb{X}$ is a Banach space.

Exercise 4.15. Let $\mathbb{X}$ be the space of bounded real functions on $[0,3]$ which (a) when restricted to $[0,1]$ are continuous, and (b) satisfy the condition

$$x(t) = \tfrac{1}{2}(x(t-1) + x(t+1)), \qquad x \in [1,2].$$

What is a typical member of this space like? Prove that $\mathbb{X}$ is a Banach space.

Exercise 4.16. Show that the space ℓ^1 of absolutely summable sequences $x = (\xi_i)_{i\geq 1}$ is a Banach space when equipped with any of the following norms:

$$\|x\| = \sup_{i=1,\ldots,1000} \frac{|\xi_i|}{i} + \sum_{i=1001}^{\infty} |\xi_i|,$$

$$\|x\| = \sum_{i=1}^{\infty} \frac{i}{i+10^{100}} |\xi_{2i}| + \sum_{i=0}^{\infty} |\xi_{2i+1}|,$$

$$\|x\| = \sum_{i=1}^{\infty} \mathrm{e}^{\sin\frac{i\pi}{7}} |\xi_i|.$$

Exercise 4.17. A function $x\colon \mathbb{R} \to \mathbb{R}$ is said to be uniformly continuous if for every $\epsilon > 0$ there is a $\delta > 0$ such that $|s-t| < \delta$ implies $|x(s) - x(t)| < \epsilon$; $\mathrm{BUC}(\mathbb{R})$ (alternatively: $\mathrm{UC}_b(\mathbb{R})$) is used to denote the space of uniformly continuous functions that are additionally bounded. It is easy to see that linear combinations of bounded uniformly continuous functions are bounded and uniformly continuous. Show that with norm $\|x\| = \sup_{t\in\mathbb{R}} |x(t)|$, BUC($\mathbb{R}$) is a Banach space. **Hint:** Modify the argument of the end of Section 3.2.

Exercise 4.18. Prove that $C[0,1]$ is not a Banach space when equipped with the norm $\|x\| = \int_0^1 |x(s)|\, \mathrm{d}s$, by examining the sequence $(x_n)_{n\geq 1}$ given by

$$x_n(s) = \begin{cases} 0, & 0 \leq s \leq \frac{1}{2} - \frac{1}{n}, \\ \frac{1}{2} + \frac{n}{2}(s - \frac{1}{2}), & \frac{1}{2} - \frac{1}{n} \leq s \leq \frac{1}{2} + \frac{1}{n}, \\ 1, & \frac{1}{2} + \frac{1}{n} \leq s \leq 1. \end{cases}$$

Exercise 4.19.

(a) Show that the subspace $\mathbb{X}$ of ℓ^1 formed by sequences such that $\sum_{i=1}^{\infty} \xi_i = 0$ with the usual ℓ^1 norm is a Banach space.
(b) Explain why replacing condition $\sum_{i=1}^{\infty} \xi_i = 0$ above by

there exists $i \geq 1$ *such that* $\xi_i = 0$,

would lead to a space that is not complete.

Exercise 4.20. Use an induction argument to check that, for each n,

$$\mathrm{e}^t = \sum_{k=0}^{n} \frac{t^k}{k!} + \int_0^t \frac{(t-s)^n}{n!} \mathrm{e}^s \, \mathrm{d}s. \tag{4.9}$$

Conclude that the sequence $(x_n)_{n\geq 1}$ of Section 4.4.2 converges in $C[0,1]$, that is, in the supremum norm, to the exponential function.

Exercise 4.21. A sequence $x = (\xi_i)_{i\geq 1}$ belongs, by definition, to the space c_{00} if there is $n = n(x)$ such that $\xi_i = 0$ for all $i \geq n$. Check to see that, when equipped with

$$\|x\| = \sup_{i\geq 1} |\xi_i|,$$

this is a normed space which, however, is not complete. (Do not forget to check that, for $(\xi_i)_{i\geq 1} \in c_{00}$, $\|(\xi_i)_{i\geq 1}\|$ is finite.) Repeat the exercise for the norms

$$\|x\| = \sup_{i\geq 1} (i+1)^2 |\xi_i|$$

and

$$\|x\| = \sum_{i=1}^{\infty} |\xi_i|.$$

Exercise 4.22. Let $C[0,2]$ be the space of real continuous functions on $[0,2]$. Check to see that, when equipped with the norm

$$\|x\| = \sup_{t\in[0,1]} |x(t)| + \int_1^2 |x(t)|\,\mathrm{d}t,$$

this is a linear normed space, but this space is not complete.

Exercise 4.23. Let $\mathbb{X}$ be the space of bounded real functions on $S = [0,2]\cup\{5\}$, such that $x(t) = 0$ for $t \in [1,2]$. Is this a Banach space when equipped with the norm $\|x\| = \sup_{t\in[0,2]}|x(t)| + |x(5)|$?

Exercise 4.24. Let $\mathbb{X}$ be the space of real continuous functions on $[0,1]$ with graphs that are symmetric about $\frac{1}{2}$ (that is, functions such that $x(t) = x(1-t)$ for $t \in [0,1]$), which additionally have the property that $x(0) = x(\frac{1}{2})$. Is this a Banach space when equipped with the norm $\|x\| = \sup_{t\in[0,1]}|x(t)|$?

Exercise 4.25. Let

$$a_i := \frac{1}{i^4 + i^2 - 1} \quad \text{and} \quad b_i = \mathrm{e}^{-4}\frac{2^i}{i!}(-1)^{i+3}, \qquad i \geq 0.$$

Show that in the space ℓ^1 of absolutely summable sequences $(\xi_i)_{i\geq 0}$ there is precisely one sequence such that

$$\xi_i + \sum_{k=0}^{i} \xi_k b_{i-k} = a_i, \qquad i \geq 0.$$

Exercise 4.26. Show that there is precisely one $(\xi_i)_{n\geq 1}$ such that

$$\sum_{i=1}^{\infty} |\xi_i| < \infty \qquad \text{and} \qquad i\xi_{i+1} = 4i^2\xi_i + 2\xi_{i+2} + \sin i, \quad i \geq 1.$$

Exercise 4.27. Show that there is precisely one $(\xi_i)_{i\in\mathbb{Z}}$ such that

$$\sum_{i\in\mathbb{Z}} |\xi_i| < \infty \quad \text{and} \quad (i^2+1)\xi_i + \tfrac{i^2\cos i}{2}\xi_{i-2} = \tfrac{(i^2+1)\sin i^2}{5}\xi_{i-1} + \tfrac{1}{8}\xi_{i+1} + 1, \quad i \in \mathbb{Z}.$$

Exercise 4.28. Use completeness of c_0 and Banach's fixed point theorem to show the following result: suppose that for a sequence $(a_i)_{i\geq 1}$ an $\alpha \in (-1,1)$ can be found such that $\lim_{i\to\infty}(a_i - \alpha a_{i-1}) = 0$. Then $(a_i)_{i\geq 1}$ is a member of c_0.

Hint: Let $(b_i)_{i\geq 1} = (a_i - \alpha a_{i-1})_{i\geq 1}$ where $a_0 := 0$, and consider the map $T\colon c_0 \to c_0$ given by $T((\xi_i)_{i\geq 1}) = S((\xi_i)_{i\geq 1}) + (b_i)_{i\geq 1}$ where S defined by $S((\xi_i)_{i\geq 1}) = (0, \xi_1, \xi_2, \dots)$ is a shift to the right. Check to see that there is precisely one fixed point of this map in c_0 and conclude, by looking at the relevant recurrence relations, that this fixed point is $(a_i)_{i\geq 1}$.

☞ CHAPTER SUMMARY

We are finally introduced to the fundamental notion of functional analysis: the Banach space, a unique blend of notions of linear algebra and metric topology. We get to know a number of classical, elementary Banach spaces. Also, examples of normed linear spaces that are not complete teach us that in a Banach space its 'extent' and its norm match each other tightly.

5
Renewal Equation in the McKendrick–von Foerster Model

Our goal in this chapter is to solve a renewal equation that is a key to the model of population dynamics due to McKendrick and von Foerster. And again completeness of a certain, well-devised space (to wit, completeness of $L^1(\mathbb{R}^+)$) is the main reason – besides the form of the renewal equation – for the existence of a unique solution. Our argument comes down to Banach's principle, but we need to start from the beginning – that is, to describe the space in which the solution will finally be found.

5.1 Banach Algebras, and the Convolution Algebra $L^1(\mathbb{R}^+)$

The space $L^1(\mathbb{R}^+)$ of (equivalence classes) of absolutely integrable functions on $\mathbb{R}^+$ is, like its kin $L^1(\mathbb{R})$, a Banach space, when equipped with the norm

$$\|\phi\| = \int_0^\infty |\phi(t)|\,\mathrm{d}t.$$

In fact, $L^1(\mathbb{R}^+)$ can be seen as a closed subspace of $L^1(\mathbb{R})$ as long as we identify a function on $\mathbb{R}^+$ with its extension to the entire $\mathbb{R}$ that vanishes on the left half-axis. Both spaces are, furthermore, examples of convolution algebras, that is, commutative Banach algebras with convolution as multiplication.

To recall, a Banach space $\mathbb{X}$ is said to be a Banach algebra if, besides the linear and topological structures discussed in the previous chapters, there is in it an additional, associative binary operation $\mathbb{X}\times\mathbb{X} \to \mathbb{X}$, called multiplication, mapping two elements, say x and y, of $\mathbb{X}$ to a third element of $\mathbb{X}$, denoted xy. This map, by definition, satisfies the following properties that tie multiplication with the Banach space:

(a) $\|xy\| \le \|x\|\,\|y\|$,
(b) $(\alpha x)y = \alpha(xy) = x(\alpha y), \alpha \in \mathbb{R}$,

(c) $x(y_1 + y_2) = xy_1 + xy_2$,
(c′) $(y_1 + y_2)x = y_1x + y_2x$.

Above, x, y, y_1 and y_2 are elements of $\mathbb{X}$, and α is a scalar. If, additionally,

(d) $xy = yx$,

we say that the algebra is commutative – in this case (c′) is redundant, as it follows from (c) and (d).

In $L^1(\mathbb{R}^+)$ as a Banach algebra, the role of multiplication is played by convolution. To be more precise, given two functions ψ_1 and $\psi_2 \in L^1(\mathbb{R}^+)$, we consider their convolution $\varphi = \psi_1 * \psi_2$ given by

$$\varphi(t) = \int_0^t \psi_1(t-s)\psi_2(s)\,\mathrm{d}s, \qquad t \geq 0. \tag{5.1}$$

The so-defined φ is a member of $L^1(\mathbb{R}^+)$ (more precisely: the function defined above is a representative of an equivalence class of functions, that is, of a member of $L^1(\mathbb{R}^+)$, and this class does not depend on the choice of representatives ψ_1 and ψ_2). This is proved by the following calculation based on the Fubini theorem:

$$\begin{aligned} \|\psi_1 * \psi_2\|_{L^1(\mathbb{R}^+)} &= \int_0^\infty \left| \int_0^t \psi_1(t-s)\psi_2(s)\,\mathrm{d}s \right| \mathrm{d}t \\ &\leq \int_0^\infty \int_s^\infty |\psi_1(t-s)\psi_2(s)|\,\mathrm{d}t\,\mathrm{d}s \\ &= \int_0^\infty \int_0^\infty |\psi_1(t)|\,|\psi_2(s)|\,\mathrm{d}t\,\mathrm{d}s \\ &= \|\psi_1\|_{L^1(\mathbb{R}^+)}\,\|\psi_2\|_{L^1(\mathbb{R}^+)}. \end{aligned} \tag{5.2}$$

In other words, if ψ_1 and ψ_2 belong to $L^1(\mathbb{R}^+)$, then so does their convolution, and

$$\|\psi_1 * \psi_2\|_{L^1(\mathbb{R}^+)} \leq \|\psi_1\|_{L^1(\mathbb{R}^+)}\,\|\psi_2\|_{L^1(\mathbb{R}^+)}.$$

This shows that condition (a) of the definition of Banach algebra is satisfied. Straightforward calculations prove that the other conditions are satisfied also (although checking associativity directly, without alluding to the Laplace transform, is a bit of a nightmare).

5.2 The McKendrick–von Foerster Model of Population Dynamics

With these preliminaries out of the way, we come to the McKendrick model.

A non-negative function ϕ defined on $[0, \infty)$ is said to be a density of an age-structured population if for all non-negative numbers $c < d$,

$$\int_c^d \phi(a)\,\mathrm{d}a$$

is the number of individuals of this population in the age between c and d. This notion should not be confused with probability density: in the case of population density the integral $\int_0^\infty \phi(a)\,\mathrm{d}a$ represents the total population size, varies in time, and rarely equals 1.

In what follows we will say that there are $\phi(a)$ individuals of age a. Of course, this is imprecise, but appeals to our intuition much better than the precise statement that there are $\int_a^{a+\delta a} \phi(c)\,\mathrm{d}c$ individuals of age between a and $a + \delta a$. Besides, the latter statement, especially in more complex sentences, is too complicated to be convenient.

The McKendrick–von Foerster model of population dynamics is an attempt to predict future population density from the basic characteristics of this population, namely birth and death rates, say

$$b = b(a) \quad \text{and} \quad \mu = \mu(a),$$

respectively, which – as indicated above – depend on a. The function $\mathrm{e}^{-\int_0^a \mu(c)\,\mathrm{d}c}$ can be thought of as the probability that an individual will reach the age $a > 0$. We assume that μ is integrable and that b is bounded; such assumptions seem to be reasonable from the biological point of view.

Let's think first of the case where $b \equiv 0$, that is, the case of no new births; here, the dynamics of the population distribution is quite simple: individuals who at time t are of age $a > t$, at time 0 were $a - t$ old. But of all those individuals who were $a - t$ old at time 0 only the fraction of $\mathrm{e}^{-\int_{a-t}^a \mu(c)\,\mathrm{d}c}$ survived till time t. This is because

$$\mathrm{e}^{-\int_{a-t}^a \mu(c)\,\mathrm{d}c} = \frac{\mathrm{e}^{-\int_0^a \mu(c)\,\mathrm{d}c}}{\mathrm{e}^{-\int_0^{a-t} \mu(c)\,\mathrm{d}c}}$$

is the probability of reaching the age a conditional on reaching the age $a - t$. In other words, if ϕ_0 is the population density at time 0 then at time t the population density has the form

$$\mathrm{e}^{-\int_{a-t}^a \mu(c)\,\mathrm{d}c}\phi_0(a - t).$$

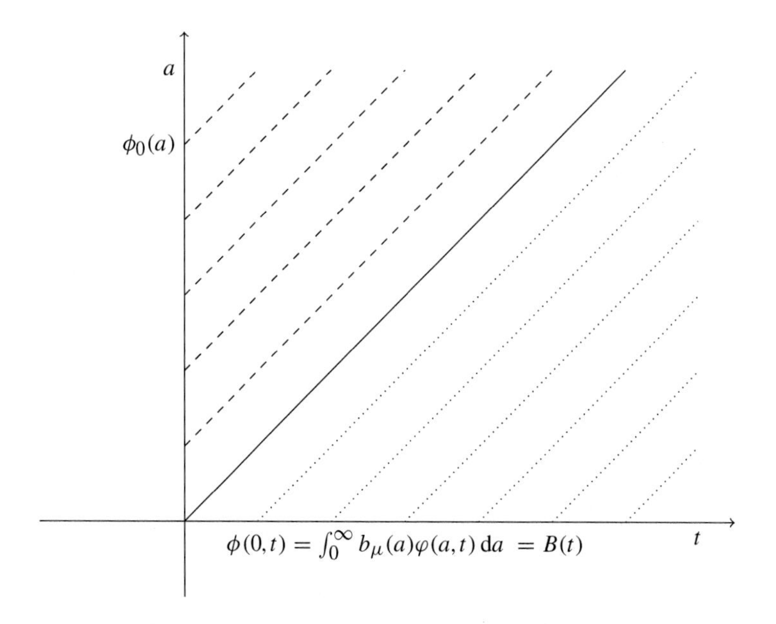

Figure 5.1 Population dynamics in the McKendrick–von Foerster model: ϕ propagates along the lines where $a - t$ is constant. In the upper quarter-space, ϕ is a function of ϕ_0. In the lower, ϕ depends on B, and B is not a priori given, but rather found as a solution to the renewal equation

On the other hand, for $a < t$, since there are no births, the population density is 0. Writing $\phi(t,\cdot)$ to denote the population density at time $t > 0$, we obtain

$$\phi(t,a) = \begin{cases} \mathrm{e}^{-\int_{a-t}^{a} \mu(c)\,\mathrm{d}c}\phi_0(a-t), & a \geq t, \\ 0, & a < t. \end{cases} \tag{5.3}$$

Of course, the situation where $b \not\equiv 0$ is more interesting and complex. To study this case, let's consider

$$B(t) := \int_0^\infty b(a)\phi(t,a)\,\mathrm{d}a, \qquad t \geq 0.$$

Intuitively, $B(t)$ is the number of individuals who are born in the entire population at time t. Were B known, we could modify formula (5.3), by replacing 0 in the second row by $\mathrm{e}^{-\int_0^a \mu(c)\,\mathrm{d}c} B(t-a)$:

$$\phi(t,a) = \begin{cases} \mathrm{e}^{-\int_{a-t}^{a} \mu(c)\,\mathrm{d}c}\phi_0(a-t), & a \geq t, \\ \mathrm{e}^{-\int_0^a \mu(c)\,\mathrm{d}c} B(t-a), & a < t \end{cases} \tag{5.4}$$

(see Figure 5.1). Indeed, each individual who at time t is of age $a < t$ must have been born a units of time ago to someone who lived at time $t - a$, and $\mathrm{e}^{-\int_0^a \mu(c)\,\mathrm{d}c}$ is the probability that such an individual survived till time t.

Our task thus reduces to finding B: as yet, (5.4) defines $\phi(t,\cdot)$ by $\phi(t,\cdot)$. However, plugging (5.4) into the definition of B, we obtain

$$B(t) = \int_0^t b(a)\mathrm{e}^{-\int_0^a \mu(c)\,\mathrm{d}c} B(t-a)\,\mathrm{d}a + \int_t^\infty b(a)\mathrm{e}^{-\int_{a-t}^a \mu(c)\,\mathrm{d}c}\phi_0(a-t)\,\mathrm{d}a \tag{5.5}$$

$$= \int_0^t \mathrm{e}^{-\int_0^{t-a} \mu(c)\,\mathrm{d}c} b(t-a)B(a)\,\mathrm{d}a + \int_0^\infty \mathrm{e}^{-\int_a^{a+t} \mu(c)\,\mathrm{d}c}\phi_0(a)b(a+t)\,\mathrm{d}a.$$

Introducing

$$C(t) = \int_0^\infty \mathrm{e}^{-\int_a^{a+t} \mu(c)\,\mathrm{d}c}\phi_0(a)b(a+t)\,\mathrm{d}a,$$

we write this formula in a more condensed way:

$$B = b_\mu * B + C, \tag{5.6}$$

where $b_\mu(t) := \mathrm{e}^{-\int_0^t \mu(c)\,\mathrm{d}c} b(t)$. It should be noted here that C depends merely on μ, b and the initial population density ϕ_0, and thus can be treated as known. Relation (5.6) is a special case of a Volterra equation of the second type, and a convolution equation at the same time. In the context of the McKendrick–von Foerster model, it is referred to as a renewal equation.

5.3 The Renewal Equation and Its Solution

What does this equation mean? There are two categories of children who are born at time t. The first of these categories is composed of children of individuals who existed at time 0, and $C(t)$ is the number of such children. Indeed, initially there were $\phi_0(a)$ individuals of age a, and $\mathrm{e}^{-\int_a^{a+t} \mu(c)\,\mathrm{d}c}\phi_0(a)$ of them survived till time $t > 0$, and are now of age $a+t$. Thus, there are $C(t)$ children born to them at time t, because $b(a+t)$ is the rate at which children are born to individuals of age $a+t$.

The second class consists of children of individuals who were born in the meantime (either to the members of the original population, or to their children, or to their grandchildren, etcetera). A parent of a child of this class must have been born before time t, and there are $B(a)$ parents born at time $a \in [0,t]$. These parents survive till time t with probability $\mathrm{e}^{-\int_0^{t-a} \mu(c)\,\mathrm{d}c}$ and at that time are of age $t-a$. It follows that the first integral in (5.5) is the total number of children of the second class.

If we agree that convolution resembles multiplication (and indeed these operations are similar – see Exercise 5.2), equation (5.6) turns out to be similar to the linear equation. Thus, its solution should be represented by the series

$$B = \sum_{n=0}^{\infty} b_\mu^{n*} * C, \tag{5.7}$$

where b_μ^{n*} is the nth convolution power of b_μ (defined recursively). In Chapter 12 it will become clear in what sense this series converges. (See also Exercise 5.4 where we apply this series to solve (5.6) explicitly in a particular case.) For now, we content ourselves with saying that there is precisely one solution to the renewal equation, and this solution determines $\phi(t,\cdot)$ of equation (5.4)

Before we prove this result, we note that it is rather unlikely that B belongs to $L^1(\mathbb{R}^+)$: b itself is merely bounded and (for example when $\mu = 0$) need not be integrable. Likewise, in general, it will be hard to check whether C is a member of $L^1(\mathbb{R}^+)$. And this is quite logical: if there are no deaths, and the birth rate is large, the population will grow rapidly, perhaps exponentially, and so will C. For non-zero μ, this rapid growth will be slowed down, restricted, by deaths, but unless we know more about the relation between μ and b it is hard to estimate the dynamics of births, that is, the speed at which the population expands.

Therefore, we search for B in a larger space. To this end, given $\omega > 0$, we define $L^1_\omega(\mathbb{R}^+)$ as the space of (equivalence classes of) functions ϕ on $\mathbb{R}^+$ such that $\int_0^\infty \mathrm{e}^{-\omega a}|\phi(a)|\,\mathrm{d}a$ is finite. When equipped with the norm equal to the integral above, $L^1_\omega(\mathbb{R}^+)$ is a Banach space (we will prove it below, but one can also argue directly).

Although it may seem strange initially, regardless of the fact that $L^1_\omega(\mathbb{R}^+)$ is apparently much more 'spacious' than $L^1(\mathbb{R}^+)$, $L^1_\omega(\mathbb{R}^+)$ is in fact 'the same as' $L^1(\mathbb{R}^+)$ – that is the key to all that follows. This situation is somewhat similar to the case of integers and even numbers. Although there are apparently more integers than even numbers, the cardinality of these sets is in fact the same. Likewise in the situation at hand: there is a one-to-one correspondence between elements of these spaces, and, additionally, the map we have in mind is linear. To be more precise: to each $\phi \in L^1_\omega(\mathbb{R}^+)$ we can assign $I\phi \in L^1(\mathbb{R}^+)$ (in the case of linear maps it is customary to write $I\phi$ instead of $I(\phi)$, see Chapter 12), given by

$$(I\phi)(t) = \mathrm{e}^{-\omega t}\phi(t), \qquad t \geq 0, \tag{5.8}$$

and linearity of I comes down to the fact that

$$I(\alpha\phi + \beta\varphi) = \alpha I\phi + \beta I\varphi$$

for all $\phi, \varphi \in L^1_\omega(\mathbb{R}^+)$ and all scalars α and β. It is amply clear that I is injective and maps $L^1_\omega(\mathbb{R}^+)$ onto $L^1(\mathbb{R}^+)$: for each $\psi \in L^1(\mathbb{R}^+)$, the function ϕ given by $\phi(t) = \mathrm{e}^{\omega t}\psi(t)$ is a member of $L^1_\omega(\mathbb{R}^+)$, and we have $I\phi = \psi$. The very definition of the norms in these spaces also tells us that

$$\|I\phi\|_{L^1(\mathbb{R}^+)} = \|\phi\|_{L^1_\omega(\mathbb{R}^+)}. \tag{5.9}$$

And because of this relation, completness of $L^1_\omega(\mathbb{R}^+)$ is a consequence of the completeness of $L^1(\mathbb{R}^+)$.

Indeed, if $(\phi_n)_{n\geq 1}$ is a Cauchy sequence in $L^1_\omega(\mathbb{R}^+)$ then its image $(I\phi_n)_{n\geq 1}$ is a Cauchy sequence in $L^1(\mathbb{R}^+)$. Thus, completness of $L^1(\mathbb{R}^+)$ implies that there is a $\psi \in L^1(\mathbb{R}^+)$ to which $(I\phi_n)_{n\geq 1}$ converges. This ψ, however, is an image of a certain $\phi \in L^1_\omega(\mathbb{R}^+)$: $\psi = I\phi$. Relation (5.9) combined with linearity of I then yields

$$\|\phi_n - \phi\|_{L^1_\omega(\mathbb{R}^+)} = \|I(\phi_n - \phi)\|_{L^1(\mathbb{R}^+)} = \|I\phi_n - I\phi\|_{L^1_\omega(\mathbb{R}^+)},$$

and the right-hand side converges to 0, as $n \to \infty$. This shows, however, that $\phi = \lim_{n\to\infty}\phi_n$, completing the proof of completeness of $L^1_\omega(\mathbb{R}^+)$.

The implications of (5.9) go much beyond completeness; this relation shows that $L^1(\mathbb{R}^+)$ and $L^1_\omega(\mathbb{R}^+)$ are in fact isomorphic, that is, have identical 'shape.' In particular, $L^1_\omega(\mathbb{R}^+)$ is a Banach algebra, since $L^1(\mathbb{R}^+)$ is. To see this, let's introduce multiplication in $L^1_\omega(\mathbb{R}^+)$, denoted, for the time being, by $\star$, by the following formula (see Figure 5.2):

$$\phi \star \varphi = I^{-1}[(I\phi) * (I\varphi)].$$

Checking that $\star$ satisfies the conditions given in the definition of a Banach algebra is quite easy; for example, condition (a) is proved as follows:

$$\begin{aligned}\|\phi \star \varphi\|_{L^1_\omega(\mathbb{R}^+)} &= \|(I\phi) * (I\varphi)\|_{L^1(\mathbb{R}^+)} \leq \|I\phi\|_{L^1(\mathbb{R}^+)}\|I\varphi\|_{L^1(\mathbb{R}^+)}\\ &= \|\phi\|_{L^1_\omega(\mathbb{R}^+)}\|\varphi\|_{L^1_\omega(\mathbb{R}^+)};\end{aligned}$$

in this calculation, we have used (5.9) (twice), and (not surprisingly) the fact that $L^1(\mathbb{R}^+)$ is an algebra. As it turns out, an explicit formula for $\star$ can be given. To this end, we note that

$$\begin{aligned}[(I\phi) * (I\varphi)](t) &= \int_0^t \mathrm{e}^{-\omega(t-s)}\phi(t-s)\mathrm{e}^{-\omega s}\varphi(s)\,\mathrm{d}s\\ &= \mathrm{e}^{-\omega t}\int_0^t \phi(t-s)\varphi(s)\,\mathrm{d}s\\ &= [I(\phi * \varphi)](t),\end{aligned}$$

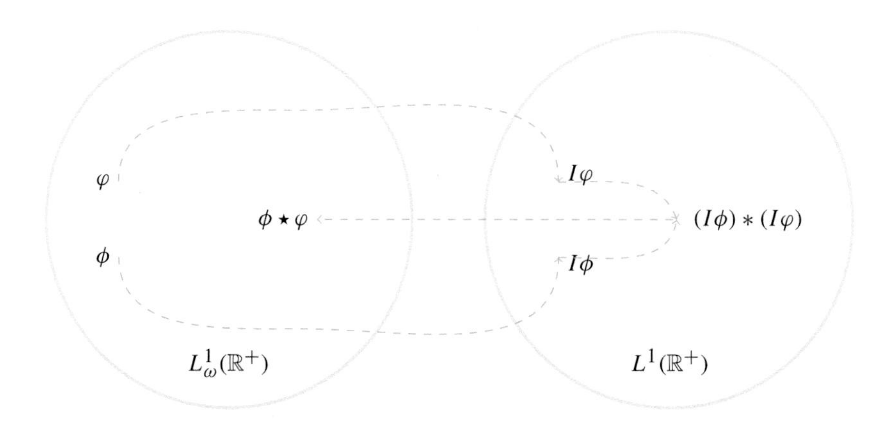

Figure 5.2 Definition of multiplication in $L^1_\omega(\mathbb{R}^+)$: find the images of ϕ and ψ in $L^1(\mathbb{R}^+)$, convolve them together, and then come back to $L^1_\omega(\mathbb{R}^+)$. The diagram shown here uses the bold hypothesis of Schar Scharikov that Banach spaces are round shaped. However, since other scientists suggest that these spaces resemble rather kidneys (especially the research group in Nierestädt, Germany), whereas others maintain that they resemble eggs (René Oeuf), our formal reasoning does not use this hypothesis.[a]

[a] Referee's comment on Figure 5.2: will the students understand this dry joke? I am afraid this can became a foundation for a new school of functional analysis.

and this shows that

$$(\phi \star \varphi)(t) = (\phi * \varphi)(t).$$

In other words, $\star$ is simply a convolution.

We are finally ready to show that equation (5.6) has precisely one solution – and that this solution belongs to $L^1_\omega(\mathbb{R}^+)$ provided ω is sufficiently large. We begin by noting that C belongs to $L^1_\omega(\mathbb{R}^+)$. Indeed,

$$\begin{aligned}\|C\|_{L^1_\omega(\mathbb{R}^+)} &\le \int_0^\infty \mathrm{e}^{-\omega t} \int_0^\infty |\phi_0(a)| b(a+t)\,\mathrm{d}a\,\mathrm{d}t \le \frac{\|b\|_\infty}{\omega} \int_0^\infty |\phi_0(a)|\,\mathrm{d}a \\ &= \frac{\|b\|_\infty}{\omega} \|\phi_0\|_{L^1(\mathbb{R}^+)} < \infty.\end{aligned}$$

Likewise with b_μ: since this function is bounded, $t \mapsto \mathrm{e}^{-\omega t} b_\mu(t)$ is integrable and

$$\|b_\mu\|_{L^1_\omega(\mathbb{R}^+)} \le \frac{\|b_\mu\|_\infty}{\omega}.$$

It follows that T defined by

$$T(\psi) = C + b_\mu * \psi, \qquad \psi \in L^1_\omega(\mathbb{R}^+), \tag{5.10}$$

maps $L^1_\omega(\mathbb{R}^+)$ into itself. Furthermore,

$$\|T(\psi_1) - T(\psi_2)\|_{L^1_\omega(\mathbb{R}^+)} = \|b_\mu * (\psi_1 - \psi_2)\|_{L^1_\omega(\mathbb{R}^+)} \le \frac{\|b\|_\infty}{\omega} \|\psi_1 - \psi_2\|_{L^1_\omega(\mathbb{R}^+)}.$$

For $\omega > \|b\|_\infty$, T is thus a contraction, establishing the existence and uniqueness of B.

5.4 Exercises

Exercise 5.1. Let $e_\lambda(t) = \mathrm{e}^{-\lambda t}, t \ge 0$, $e_\lambda \in L^1(\mathbb{R}^+), \lambda > 0$. Prove that the following Hilbert relation holds:

$$e_\lambda - e_\mu = (\mu - \lambda) e_\lambda * e_\mu.$$

Exercise 5.2. Check that $\phi * \psi = \psi * \phi, (a\phi + b\psi) * \varphi = a\phi * \varphi + b\psi * \varphi$ for all $\phi, \psi, \varphi \in L^1(\mathbb{R}^+)$ and scalars a and b. Show also that for the function i defined by $i(t) = 1, t \ge 0$ we have $i * \phi(t) = \int_0^t \phi(s)\,\mathrm{d}s$.

Exercise 5.3. Provide a recursive definition of the nth convolution power of a $\phi \in L^1(\mathbb{R}^+)$. Check to see that, for any $\lambda > 0$, the nth convolution power of ϕ defined by $\phi(t) = \lambda \mathrm{e}^{-\lambda t}, t \ge 0$ is

$$\phi^{*n}(t) = \mathrm{e}^{-\lambda t} \frac{\lambda^n t^n}{(n-1)!}, \qquad t \ge 0.$$

Exercise 5.4. Consider the particular case of McKendrick–von Foerster model with μ and b independent of a, and suppose that $\mu = b = \lambda$ where $\lambda > 0$ is given. Check to see that in this case

$$C(t) = c\lambda \mathrm{e}^{-\lambda t}, \qquad b_\mu(t) = \lambda \mathrm{e}^{-\lambda t}, \qquad t \ge 0,$$

for $c = \|\phi_0\| = \int_0^\infty \phi_0(a)\,\mathrm{d}a$. Use the previous problem to show that the formal series $\sum_{n=1}^\infty b_\mu^{*n}$ reduces then to the constant function λ. Conclude from (5.7) that the solution to equation (5.6) is given by $B(t) = c\lambda$, and then check that this function indeed solves this equation.

Exercise 5.5. Prove that ℓ^1 is a convolution algebra with convolution

$$(\xi_i)_{i \ge 0} * (\eta_i)_{i \ge 0} = (\zeta_i)_{i \ge 0},$$

where $\zeta_i = \sum_{k=0}^i \xi_{i-k} \eta_k$.

Exercise 5.6. Let $\alpha \in (0,1)$ be given. We define $a \in \ell^1$ by

$$a = (1, \alpha, \alpha^2, \cdots).$$

Prove that for each $y \in \ell^1$ there is precisely one $x \in \ell^1$ such that

$$x - \alpha(1-\alpha)a * x = y.$$

Exercise 5.7. Let multiplication in $C[a,b]$ be given by

$$(xy)(t) = x(t)y(t), \qquad t \in [a,b].$$

Prove that $C[a,b]$ with this multiplication is a Banach algebra.

Exercise 5.8. Prove that $L^1(\mathbb{R})$ is a commutative Banach algebra with convolution (as multiplication) defined by

$$(\phi * \varphi)(t) = \int_{-\infty}^{\infty} \phi(t-s)\varphi(s)\,\mathrm{d}s, \qquad t \in \mathbb{R}.$$

Check to see that if ϕ and φ vanish on the negative half-axis, their convolution, as defined above, coincides with (5.1).

Exercise 5.9. We introduce a new multiplication in $L^1(\mathbb{R}^+)$ by

$$(\phi \diamond \varphi)(t) = \tfrac{1}{2}\int_{-\infty}^{\infty} \phi(|t-s|)\varphi(|s|)\,\mathrm{d}s, \qquad t \geq 0.$$

Prove that $L^1(\mathbb{R}^+)$ is a commutative Banach algebra with this multiplication. (We should refrain from calling it a convolution algebra even though the right-hand side here is a constant multiple of the convolution of even extensions of ϕ and ψ.)

☞ CHAPTER SUMMARY

If equipped with an additional operation of multiplication of vectors by other vectors (to give yet other vectors), an operation that is well intertwined with the existing linear and topological structures, a Banach space becomes a Banach algebra. Such an additional operation is naturally defined in the space of continuous functions by pointwise multiplication. In the space of integrable functions on $\mathbb{R}^+$, however, the role of multiplication is most naturally played by convolution. We use this additional algebraic structure in $L^1(\mathbb{R}^+)$ to study the McKendrick–von Foerster model of population dynamics. Existence and uniqueness of the renewal equation that is a key to the model turns out – surprise, surprise! – to be the result of the completeness of $L^1(\mathbb{R}^+)$.

6

Riemann Integral for Vector-Valued Functions

6.1 Riemann Integral in a Normed Space

There is no modern analysis without the notion of integrals for vector-valued functions. And it is a fundamental consequence of completeness that many important results of real (and complex) calculus extend, with appropriate but minor changes, to functions with values in Banach spaces. Among many possible notions of integration, the Riemann type is arguably the most common and perhaps the simplest (at least in comparison to Lebesgue-type integrals, termed Bochner integrals, see e.g., [22]).

In this chapter we discuss two basic facts related to the theory of Riemann integration. First of all, the linear structure allows us to define approximating sums of a Riemann integral of a vector-valued function, whereas the notion of the norm allows us to define the integral as a limit of these sums. However, without additional information on the space and/or the function, the existence of the limit (which is to be independent of the way approximating sums are chosen) is not obvious. Secondly, it is completeness of a normed space that guarantees existence of the Riemann integral of a continuous, vector-valued function.

Here are the details. Let $a < b$ be real numbers and let $x\colon [a,b] \to \mathbb{X}$, $t \mapsto x(t)$ be a function on the interval $[a,b]$ taking values in a normed space $\mathbb{X}$. Let's consider two sequences $\mathcal{T} = (t_i)_{i=0,..,k}$ and $\mathcal{P} = (p_i)_{i=0,...,k-1}$ of points of $[a,b]$ (k is a natural number) that are chosen so that

$$a = t_0 < t_1 < \cdots < t_k = b, \qquad t_0 \le p_0 \le t_1 \le \cdots \le t_{k-1} \le p_{k-1} \le t_k. \tag{6.1}$$

We define the related number $\Delta(\mathcal{T}) = \sup_{0<i\le k}(t_i - t_{i-1})$ and a member of $\mathbb{X}$ given by

$$S(\mathcal{T},\mathcal{P},x) = \sum_{i=0}^{k-1}(t_{i+1} - t_i)x(p_i). \tag{6.2}$$

(To repeat, this formula makes sense because x has values in a vector space, where addition of vectors and multiplication by scalars is possible.) A function x is said to be integrable if the limit

$$\lim_{n\to\infty} S(\mathcal{T}_n,\mathcal{P}_n,x)$$

exists for all sequences of pairs $(\mathcal{T}_n,\mathcal{P}_n)$ such that $\lim_{n\to\infty}\Delta(\mathcal{T}_n) = 0$, and this limit does not depend on the choice of $(\mathcal{T}_n,\mathcal{P}_n)$. Moreover, the limit spoken of here is termed the Riemann integral of x and denoted $\int_a^b x(t)\,\mathrm{d}t$. We stress again that one cannot speak of a limit if there is no topology in $\mathbb{X}$, and in normed spaces topology is provided by the norm.

In the next section, we show that for functions taking values in Banach spaces the following theorem, known from the calculus of real-valued functions, is true: **continuous functions are Riemann integrable.** It cannot be overstressed that for functions with values in a space that is not complete, the theorem is no longer true.

6.2 Integrability of Continuous Functions

To begin the proof, let $\epsilon > 0$ be given. Since the interval $[a,b]$ is a compact set, continuity of x implies its uniform continuity. There is thus a $\delta > 0$ such that combined conditions $|s-t| < \delta$ and $s,t \in [a,b]$ force $\|x(s) - x(t)\| < \epsilon$.

Suppose now that the sequences $\mathcal{T} = (t_i)_{i=0,\ldots,k}$ and $\mathcal{T}' = (t_i')_{i=0,\ldots,k'}$ are such that $\Delta(\mathcal{T}) < \delta$ and $\Delta(\mathcal{T}') < \delta$. Suppose also that $\mathcal{P}$ and $\mathcal{P}'$ are related sequences of midpoints (so that we have (6.1)). Let $\mathcal{T}'' = (t_i'')_{i=1,\ldots,k''}$ be the sequence that contains all the members of $\mathcal{T}$ and $\mathcal{T}'$. Let also $\mathcal{P}'' = (p_i'')_{i=0,\ldots,k''-1} \equiv (t_i'')_{i=0,\ldots,k''-1}$ (so that the midpoints lie at the ends of intervals).

It is a key to the proof that

$$\|S(\mathcal{T},\mathcal{P},x) - S(\mathcal{T}'',\mathcal{P}'',x)\| \le \epsilon(b-a). \tag{6.3}$$

To establish this relation, we note that $\Delta(\mathcal{T}'') < \delta$ and $k'' \le k+k'-2$, because besides $t_0 = t_0' = a$ and $t_k = t_{k'}' = b$ there could be some other $t_i = t_j'$, $i = 1,\ldots,k-1, j = 1,\ldots,k'-1$. The interval $[t_i,t_{i+1}]$, $i = 0,\ldots,k-1$ either coincides with a certain $[t_j'',t_{j+1}'']$, $j \in \{0,\ldots,k''-1\}$ or is a sum of intervals of the latter form: suppose that $[t_i,t_{i+1}] = [t_j'',t_{j+1}''] \cup \cdots \cup [t_{j+l}'',t_{j+l+1}'']$ for certain l. Therefore,

$$\begin{aligned}
&\left\| (t_{i+1} - t_i)x(p_i) - \sum_{m=0}^{l} (t''_{j+m+1} - t''_{j+m})x(t''_{j+m}) \right\| \\
&= \left\| \sum_{m=0}^{l} (t''_{j+m+1} - t''_{j+m})[x(p_i) - x(t''_{j+m})] \right\| \\
&\le \epsilon \sum_{m=0}^{l} (t''_{j+m+1} - t''_{j+m}) = \epsilon (t_{i+1} - t_i),
\end{aligned}$$

because both p_i and t''_{j+m} belong to $[t_i, t_{i+1}]$, forcing $|p_i - t''_{j+m}| < \delta$. Summing over i, we obtain (6.3).

This reasoning works in the same way for $\mathcal{T}'$ as it does for $\mathcal{T}$. This shows that

$$\|S(\mathcal{T}, \mathcal{P}, x) - S(\mathcal{T}', \mathcal{P}', x)\| \le 2\epsilon(b - a), \tag{6.4}$$

for all $\mathcal{T}$ and $\mathcal{T}'$ satisfying conditions $\Delta(\mathcal{T}) < \delta$ and $\Delta(\mathcal{T}') < \delta$, and any sequences $\mathcal{P}$ and $\mathcal{P}'$ of midpoints. This convinces us that for any sequence $(\mathcal{T}_n, \mathcal{P}_n)$ such that $\lim_{n\to\infty} \Delta(\mathcal{T}_n) = 0$, $S(\mathcal{T}_n, \mathcal{P}_n, x)$ is fundamental. Since x has values in a complete space, $S(\mathcal{T}_n, \mathcal{P}_n, x)$ has a limit, and – because of (6.4) – this limit does not depend on the choice of $(\mathcal{T}_n, \mathcal{P}_n)$. This completes the proof.

6.3 Fundamental Theorem of Calculus

Here is another theorem of real analysis that remains the same in the calculus of vector-valued functions.

It says that if $[a,b] \ni t \to x(t) \in \mathbb{X}$ (where $\mathbb{X}$ is a Banach space) is continuous, then $[a,b] \ni t \mapsto y(t) = \int_a^t x(s)\,\mathrm{d}s$ is differentiable with $y'(t) = x(t), t \in [a,b]$ (at $t = a$ and $t = b$ we think of right-hand and left-hand derivatives, respectively).[1] For the proof we note that the difference quotient of y at $t \in (a,b)$, that is,

$$h^{-1}(y(t+h) - y(t)) = h^{-1} \int_t^{t+h} x(s)\,\mathrm{d}s,$$

where, for $h < 0$, by convention, $\int_t^{t+h} x(s)\,\mathrm{d}s := -\int_{t+h}^{t} x(s)\,\mathrm{d}s$, is well defined for h sufficiently small, and a similar statement is true for $t \in \{a,b\}$, except that for $t = a$, we must consider merely $h > 0$ and for $t = b$ only $h < 0$. We are thus left with showing that

[1] As in real analysis, a function x is differentiable at t if its difference quotients $h^{-1}(x(t+h) - x(t))$ converge as $h \to 0$; right-hand and left-hand derivatives are defined accordingly.

$$\lim_{h\to 0} h^{-1}\int_t^{t+h} x(s)\,\mathrm{d}s = x(t), \qquad \text{for all } t \in [a,b], \tag{6.5}$$

with the proviso that for $t = a$, h converges to 0 from the right, and for $t = b$, it converges to 0 from the left. This formula holds because, given $\epsilon > 0$ we can find an $h_0 > 0$ such that $\|x(s) - x(t)\| < \epsilon$ as long as $|t - s| < h_0$ and $s \in [a,b]$, and then, by (6.8),

$$\begin{aligned}\left\| h^{-1}\int_t^{t+h} x(s)\,\mathrm{d}s - x(t)\right\| &= |h^{-1}| \left\| \int_t^{t+h} [x(s) - x(t)]\,\mathrm{d}s \right\| \\ &\le |h^{-1}| \left| \int_t^{t+h} \|x(s) - x(t)\|\,\mathrm{d}s \right| \\ &< |h^{-1}| \left| \int_t^{t+h} \epsilon\,\mathrm{d}s \right| = \epsilon,\end{aligned}$$

as long as $|h| < h_0$.

To summarize, we have proved the first fundamental theorem of calculus, saying that

$$\frac{\mathrm{d}}{\mathrm{d}t}\int_a^t x(s)\,\mathrm{d}s = x(t), \qquad t \in [a,b]. \tag{6.6}$$

The second fundamental theorem of calculus deals with the case where the order of integration and differentiation in (6.6) is reversed. In our case it says that if $x'(s)$ exists for all $s \in [a,b]$ and depends continuously on s, then

$$\int_a^t x'(s)\,\mathrm{d}s = x(t) - x(a), t \in [a,b]. \tag{6.7}$$

The proof is quite simple, if we know that a vector-valued function that is differentiable with derivative equal to zero everywhere is constant. For, assuming this result, we consider

$$y(t) = x(t) - \int_a^t x'(s)\,\mathrm{d}s, \qquad t \in [a,b].$$

By (6.6), y is differentiable with $y'(t) = x'(t) - x'(t) = 0$, and thus constant. Since $y(a) = x(a)$, we have $y(t) = x(a), t \in [a,b]$, but this implies (6.7).

The shortest path to the fact that a vector-valued function that is differentiable with derivative equal to zero everywhere is constant, leads through the Hahn–Banach extension theorem, one of the fundamental results of functional analysis that, unfortunately, is not proved in this book. We discuss the theorem briefly in the Appendix, and it is to the Appendix that we refer the interested reader for the proof of the fact just stated.

6.4 Exercises

Exercise 6.1. Let $\mathbb{X}$ be a Banach space, and suppose that $t \mapsto x(t) \in \mathbb{X}$ is continuous on an interval $[a,b]$. The real-valued function $t \mapsto \|x(t)\|$ is then continuous also (because of the triangle inequality) and is thus integrable. Prove that (cf. Exercise 3.13)

$$\left\| \int_a^b x(t)\,\mathrm{d}t \right\| \leq \int_a^b \|x(t)\|\,\mathrm{d}t. \tag{6.8}$$

Exercise 6.2. Let $(a_i)_{i\geq 1}$ be a sequence of positive numbers with $\lim_{i\to\infty} a_i = a > 0$, and let $x\colon [0,1] \to c$ be given by $x(t) = \left(\mathrm{e}^{-a_i t}\right)_{i\geq 1}, t \in [0,1]$. Prove that x is continuous and find $\int_0^1 x(t)\,\mathrm{d}t \in c$. Write inequality (6.8) for this particular function.

Exercise 6.3. Let $t \mapsto u(t)$ be a continuous real-valued function defined on an interval $[a,b]$, and let x be a member of a normed space $\mathbb{X}$. Check to see that the function $t \mapsto u(t)x \in \mathbb{X}$ is Riemann integrable and we have

$$\int_a^b u(t)x\,\mathrm{d}t = \left(\int_a^b u(t)\,\mathrm{d}t\right)x.$$

(We do not assume here that $\mathbb{X}$ is complete).

Exercise 6.4. Prove that, in the set-up of Section 6.3,

$$\frac{\mathrm{d}}{\mathrm{d}t}\int_t^b x(s)\,\mathrm{d}s = -x(t), \qquad t \in [a,b].$$

☞ CHAPTER SUMMARY

The Riemann integral for vector-valued functions can be defined in the same way as for scalar-valued functions. Moreover, the theory of the so-defined integral is rather similar to the classical one. In particular, any continuous function with values in a Banach space is Riemann integrable, and the fundamental theorem of calculus remains valid. The theory of Riemann integration for functions with values in a normed, not complete, space would not be so elegant.

7
The Stone–Weierstrass Theorem

7.1 The Main Theorem

The main subject of this chapter is the following theorem of Weierstrass, a theorem to which we will come back several times in this little book.

7.1 Theorem (Weierstrass) *Any continuous (real-valued) function on the closed interval $[a,b]$ can be uniformly approximated by polynomials. In other words, for any $x \in C[a,b]$ and any $\epsilon > 0$, a polynomial p can be found so that*

$$\|x - p\| = \sup_{s \in [a,b]} |x(s) - p(s)| < \epsilon.$$

In fact, we will be able to prove a more general theorem, the Stone–Weierstrass theorem, that applies to the situation where the continuous function spoken of above instead of being defined on $[a,b]$ is defined on an abstract compact topological (Hausdorff) space S. Of course, in such generality, it makes no sense to speak of polynomials, and therefore we will discuss abstract properties of the space of polynomials that lie behind the fact that this space is dense in $C[a,b]$. The reader will note that we make a typical maneuver for functional analysis: we are not interested in polynomials themselves or their individual properties; rather, we are interested in the properties of the space of polynomials as a whole.

Before continuing, let's recall some notions of basic topology. To begin with, a topological space S is said to be compact if any cover of S contains a finite sub-cover: to be more precise, if $\mathcal{U}_\iota, \iota \in I$ (I is a set of indexes) are open sets such that $\bigcup_{\iota \in I} \mathcal{U}_\iota = S$ then there is a finite subset J of I such that $\bigcup_{\iota \in J} \mathcal{U}_\iota = S$. A real-valued function x on a topological space S is said to be continuous if inverse images (via this function) of open subsets of $\mathbb{R}$ are open in S; in symbols: $x^{-1}(\mathcal{U})$ is open in S whenever $\mathcal{U}$ is open in $\mathbb{R}$. It is a well-known fact that a continuous function on a compact set attains a maximum and a minimum. In particular, if $C(S)$ denotes the space of continuous functions on a compact space S, for each $x \in C(S)$ we can think of

$$\|x\| = \sup_{s \in S} |x(s)|.$$

Similarly as in the case of an interval we argue that the so-defined $\|\cdot\|$ is indeed a norm, and that $C(S)$ with this norm is a Banach space.

We are ready to discuss the key properties of the space P of polynomials $P \subset C[a,b]$. First of all,

(a) if $s \neq t$ are points in $[a,b]$, then there is a $p \in P$ such that $p(s) \neq p(t)$.

This property is often described by saying that P *separates points* of $[a,b]$. It is perhaps worth noting that in fact there are many polynomials that satisfy (a), and among them there is the linear function with appropriately chosen coefficients:

$$p(r) = ar + b = \frac{1}{t-s}r - \frac{s}{t-s}, \qquad r \in [a,b].$$

The second key property of the set of polynomials is even simpler than the first:

(b) for each p and q in P, their product pq is also a polynomial, and so is their linear combination $\alpha p + \beta q$, for arbitrary scalars α and β.

This property is easy to check (for instance by induction on the degree of one of the polynomials involved), but it is hard to overestimate its importance; this algebraic (it is algebraic, isn't it?) property is a key to the theorem we want to discuss here.

The theorem, the Stone–Weierstrass theorem, involves a subset P of the space $C(S)$. Even though it is still denoted by P, it is no longer a set of polynomials – as we have mentioned before, it is even unclear what the word 'polynomial' could stand for in this context. However, P is a proper counterpart of the set of polynomials, because (by assumption) it possesses the properties that are characteristic of the set of polynomials. It possesses the properties that cause it to be dense in $C(S)$. More specifically,

(A) P separates points of S, that is, given two distinct points, s and t, of S, one can find a $p \in P$ such that $p(s) \neq p(t)$, and
(B) P is a sub-algebra of $C(S)$, that is, a product and any linear combination of two elements of P belong to P.

Clearly, (A) and (B) are abstract reflections of properties (a) and (b) of the set of polynomials.

We are finally ready to present the theorem this chapter is devoted to.

7.2 Theorem (Stone–Weierstrass) *Any sub-algebra of $C(S)$ that separates points of S and contains constant functions is dense in $C(S)$.*

Since the closure of a sub-algebra that separates points is a sub-algebra that separates points, the Stone–Weierstrass theorem can be stated equivalently as follows.

7.3 Theorem (Stone–Weierstrass) *Each* closed *sub-algebra P of $C(S)$ that separates points of S coincides with $C(S)$ provided it contains constant functions.*

7.2 Proof

The following proof of the Stone–Weierstrass theorem is due to J. Zemánek[1]. There are, of course, dozens of proofs of this theorem; we have chosen the one published in [43] because it stresses the role of completeness of the space $C(S)$. We start with a lemma.

7.4 Lemma *Suppose that in a commutative Banach algebra $\mathbb{X}$ there is an element u, termed unity, such that*

$$ux = xu = x \qquad \textit{for all } x \in \mathbb{X}.$$

Furthermore, let $\epsilon \in (0,1)$ be a given number, and x be a vector such that $\|x - u\| \le \epsilon$. Then, there is a $y \in \mathbb{X}$ such that

$$\|y\| \le \epsilon \qquad \textit{and} \qquad x = (u - y)^2.$$

Proof The closed ball $K \subset \mathbb{X}$ with center at 0 and radius ϵ is, by definition, the set of all x with $\|x\| \le \epsilon$. Since K is a closed subset of a complete space, K itself is a complete metric space. Let's think of the map $T\colon K \to K$ given by

$$Tz = \frac{1}{2}\left(u - x + z^2\right)$$

(we have $\|Tz\| \le \frac{1}{2}(\epsilon + \epsilon^2) \le \epsilon$). The following calculation shows that T is a contraction with constant $\epsilon < 1$:

[1] J. Zemánek (1946–2017) was a Czech mathematician, who for most of his life was affiliated with the Institute of Mathematics of the Polish Academy of Sciences in Warsaw, Poland.

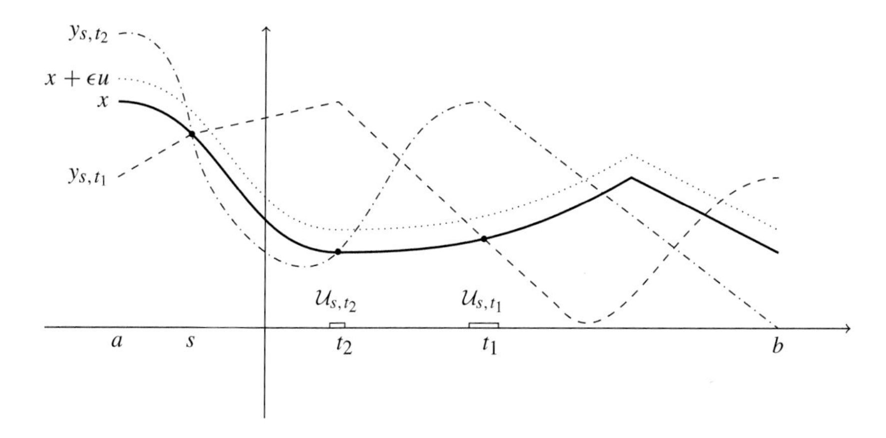

Figure 7.1 At s and t, $y_{s,t}$ has the same value as x. Since both x and $y_{s,t}$ are continuous, $y_{s,t}$ is smaller than $x + \epsilon u$ in a small neighborhood $\mathcal{U}_{s,t}$ of t. A finite number of such neighborhoods covers S (here depicted as the interval $[a, b]$), hence the minimum of corresponding functions $y_{s,t}$ (s remains fixed) lies below $x + \epsilon u$ in the entire S. The figure makes it clear that the minimum of y_{s,t_1} and y_{s,t_2} lies below $x + \epsilon u$ at least in the union of $\mathcal{U}_{s,t_1}$ and $\mathcal{U}_{s,t_2}$, whereas its value at s still equals $x(s)$.

$$\|Tz_1 - Tz_2\| = \left\| \frac{z_1^2 - z_2^2}{2} \right\| \le \left\| \frac{z_1 + z_2}{2} \right\| \|z_1 - z_2\| \le \epsilon \|z_1 - z_2\|.$$

(In the second step use the identity $z_1^2 - z_2^2 = (z_1 - z_2)(z_1 + z_2)$, which is a consequence of the commutativity of $\mathbb{X}$.) The Banach fixed point theorem tells us that there is a $y \in K$ such that

$$2y = u - x + y^2.$$

In other words, $x = u - 2y + y^2 = (u - y)^2$, completing the proof. □

7.5 Remark The space $C(S)$ possesses the unit u, defined by $u(s) = 1, s \in S$, but $L^1(\mathbb{R}^+)$ is an algebra without a unit.

Besides Zemánek's lemma, presented above, we need the following result that is of its own interest and importance, and in fact deserves to be called a theorem. The only reason why we call it a lemma here is that in the context considered it is of auxiliary character.

7.6 Lemma (Kakutani–Krein Theorem) *Let $V \subset C(S)$ be a subspace (in the linear algebra sense) of $C(S)$ that separates points and contains the function, denoted u, that equals 1 for all $s \in S$. If V is a lattice, then it is dense in $C(S)$.*

Proof To recall, $V \subset C(S)$ is said to be a lattice if together with $x, y \in V$ it contains the functions $x \vee y$ and $x \wedge y$ defined as follows:

$$(x \vee y)(s) = \max(x(s), y(s)), \qquad (x \wedge y)(s) = \min(x(s), y(s)).$$

Let $x \in C(S)$ be fixed. Our goal is, for each $\epsilon > 0$, to find a $z \in V$ such that $\|x - z\| < \epsilon$.

To this end, given $s, t \in S$ we choose a $y \in V$ so that $y(s) \neq y(t)$. Then there are unique scalars α and β such that $\alpha + \beta y(s) = x(s)$ and $\alpha + \beta y(t) = x(t)$, because the determinant of this linear system of equations for α and β is non-zero. In this manner we obtain the function $y_{s,t} := \alpha u + \beta y$, which belongs to V and at s and t coincides with x.

Next, we choose $\epsilon > 0$ and fix $s \in S$ (see Figure 7.1). Since the functions $y_{s,t}$ and x are continuous, there is an open neighborhood $\mathcal{U}_{s,t}$ of t such that $y_{s,t}(r) < x(r) + \epsilon$ for $r \in \mathcal{U}_{s,t}$. These neighborhoods, indexed by t, cover S. Because of compactness of S, there is a finite subcover of this cover. Let $t_1, ..., t_n$ be the points that determine this subcover. The function $y_s := y_{s,t_1} \wedge y_{s,t_2} \wedge \cdots \wedge y_{s,t_n}$ belongs by assumption to V and satisfies $y_s < x + \epsilon u$. Indeed, each $r \in S$ belongs to at least one of neighborhoods $\mathcal{U}_{s,t_i}$; hence, for appropriate i, $y_s(r) \leq y_{s,t_i}(r) < x(r) + \epsilon$. Moreover, we still have $y_s(s) = x(s)$.

Next, let $\mathcal{V}_s$ be an open neighborhood of s such that $y_s(r) > x(r) - \epsilon$ for $r \in \mathcal{V}_s$. The neighborhoods $\mathcal{V}_s, s \in S$ cover S. Let $s_1, ..., s_k \in S$ be such that $\mathcal{V}_{s_i}, i = 1, \ldots, k$ cover S. The function $z := y_{s_1} \vee y_{s_2} \vee \cdots \vee y_{s_n}$ belongs to V, and arguing as above we see that $z > x - \epsilon u$. Since obviously $z < x + \epsilon u$, we conclude that $\|x - z\| < \epsilon$. This completes the proof, ϵ being arbitrary. □

It should perhaps be stressed here that the set of polynomials is not a lattice, as can be seen from the fact that, for example, the minimum of $p_1, p_2 \in C[-1,1]$ defined by $p_1(s) = s$ and $p_2(s) = s^2, s \in [-1,1]$, is not a polynomial. Hence, there is no reason to expect that the algebra of the Stone–Weierstrass theorem is a lattice either. The rest of the proof of the theorem hinges on the fact that the closure of this algebra *is* a lattice.

Proof of the Stone–Weierstrass Theorem 7.2 Let V be the closure of P. Our first goal is to show that together with x, V also contains $|x|$ (the latter function is defined by $|x|(s) = |x(s)|$). Obviously, we may restrict our attention to $x \neq 0$. To this end, we consider

$$x_n(s) := \frac{1}{\|x^2\| + \frac{1}{n}} \left([x(s)]^2 + \frac{1}{n} \right), \qquad s \in S, n \geq 1.$$

The so-defined functions are continuous, positive and of norm 1. Additionally,

$$x_n \geq \delta_n u, \qquad \text{where} \qquad \delta_n := \frac{\frac{1}{n}}{\|x^2\| + \frac{1}{n}} \in (0,1).$$

It follows that the distance $\|u - x_n\|$ of x_n from u is at most $\epsilon_n := 1 - \delta_n \in (0,1)$. Since x_n are positive, we can consider $\sqrt{x_n}$, and Lemma 7.4 tells us that these new functions are elements of V (here, we use the fact that for a positive number a there is precisely one positive b with $b^2 = a$). On the other hand, for any real number a we have

$$\sqrt{a^2 + \frac{1}{n}} - |a| = \frac{\frac{1}{n}}{\sqrt{a^2 + \frac{1}{n}} + |a|} \leq \frac{\frac{1}{n}}{\sqrt{\frac{1}{n}}} = \frac{1}{\sqrt{n}}.$$

This shows that

$$\left\| \sqrt{x^2 + \frac{1}{n}u} - |x| \right\| = \sup_{s \in S} \left| \sqrt{x^2(s) + \frac{1}{n}} - |x(s)| \right| \leq \frac{1}{\sqrt{n}},$$

and, consequently, $\left\| \sqrt{x_n} - \frac{|x|}{\sqrt{\|x^2\| + \frac{1}{n}}} \right\|$ converges to 0, as $n \to \infty$. It follows that $\frac{1}{\sqrt{\|x^2\|}}|x|$ belongs to V, V being closed, and thus so does $|x|$.

Finally, we note the following formulae:

$$x \vee y = \frac{1}{2}(x + y + |x - y|) \qquad \text{and} \qquad x \wedge y = \frac{1}{2}(x + y - |x - y|), \tag{7.1}$$

which can be easily checked to hold for any x and y. Combining them with the already established facts and with the assumption that V is a linear subspace of $C(S)$, we conclude that together with x and y, V contains $x \vee y$ and $x \wedge y$. This means, however, that V is a lattice. The Kakutani–Krein theorem tells us thus that V is dense in $C(S)$. On the other hand, V is closed, and so we must have $V = C(S)$. □

We complete this chapter with two chosen applications of the Stone–Weierstrass theorem.

7.7 Corollary (Hausdorff Moment Problem) There are no other $x \in C[a,b]$ such that

$$\int_a^b x(s)s^n \, \mathrm{d}s = 0, \qquad n \geq 0, \tag{7.2}$$

except for $x = 0$.

Proof Suppose x satisfies (7.2). Given $\epsilon > 0$ we find a polynomial p such that $\|x - p\| < \epsilon$. By (7.2), we have $\int_a^b x(s)p(s)\,\mathrm{d}s = 0$. Since

$$\int_a^b x^2(s)\,\mathrm{d}s = \int_a^b x(s)[x(s) - p(s)]\,\mathrm{d}s + \int_a^b x(s)p(s)\,\mathrm{d}s = \int_a^b x(s)[x(s) - p(s)]\,\mathrm{d}s$$
$$\leq (b-a)\|x\|\,\|x - p\| < \epsilon(b-a)\|x\|,$$

and $\epsilon > 0$ is arbitrary, we conclude that $\int_a^b x^2(s)\,\mathrm{d}s = 0$. But this is impossible unless $x = 0$. □

7.8 Corollary (Trigonometric Polynomials) Let $C_{\mathrm{p}}[-\pi,\pi]$ be the space of continuous functions $x \in C[-\pi,\pi]$ such that $x(-\pi) = x(\pi)$ ('p' for 'periodic'). Any member of this space may be identified with a continuous function on a unit circle on a plane if $t \in [-\pi,\pi]$ is thought of as the angle between the x-axis and the vector connecting the center with a point on the circle. In other words, $[-\pi,\pi]$ with identified points $-\pi$ and π is a complete, compact metric space with distance[2] $d(s,t) = \min\{|s-t|, 2\pi - |t-s|\}$. An $x \in C_{\mathrm{p}}[-\pi,\pi]$ that is of the form

$$x(t) = a_0 + \sum_{\ell=1}^{n}(a_\ell \cos \ell t + b_\ell \sin \ell t)$$

for some integer n and numbers $a_0,\ldots,a_n$ and $b_1,\ldots,b_n$, is said to be a trigonometric polynomial. We will show that the set P of trigonometric polynomials is dense in $C_{\mathrm{p}}[-\pi,\pi]$, by checking that the conditions listed in the Stone–Weierstrass theorem are satisfied. First of all, P contains the constant function. Second, it separates points. Indeed, conditions $\cos s = \cos t$ and $\sin s = \sin t$ imply $s = t$; thus, for $s \neq t$, we either have $\sin s \neq \sin t$ or $\cos s \neq \cos t$. Finally, the well-known identities

$$\begin{aligned}\cos nt \cos mt &= \tfrac{1}{2}[\cos(n+m)t + \cos(n-m)t],\\ \sin nt \sin mt &= \tfrac{1}{2}[\cos(n-m)t - \cos(n+m)t],\\ \cos nt \sin mt &= \tfrac{1}{2}[\sin(n+m)t + \sin(n-m)t]\end{aligned} \tag{7.3}$$

show that P is a sub-algebra of $C_{\mathrm{p}}[-\pi,\pi]$.

7.3 Exercises

Exercise 7.1. Check to see that the closure of a sub-algebra is a sub-algebra.

Exercise 7.2. Prove relations (7.1).

[2] The distance between two points on a circle is the length of the shorter arc connecting them.

Exercise 7.3. Prove relations (7.3) using the following Euler identities:

$$\cos\alpha = \frac{1}{2}(\mathrm{e}^{i\alpha} + \mathrm{e}^{-i\alpha}) \qquad \text{and} \qquad \sin\alpha = \frac{1}{2i}(\mathrm{e}^{i\alpha} - \mathrm{e}^{-i\alpha}). \tag{7.4}$$

Exercise 7.4. Let $C_{\mathrm{e}}[-1,1]$ be the space of even continuous functions on $[-1,1]$. Prove that the space of polynomials that involve only even powers is dense in $C_{\mathrm{e}}[-1,1]$.

Exercise 7.5. Let $C[0,\infty]$ be the space of continuous functions on the right half-axis $[0,\infty)$ with finite limits at ∞; we equip the space with the norm $\|x\| := \sup_{t\geq 0}|x(t)|$. Also, for $\lambda \geq 0$, let $e_\lambda(t) = \mathrm{e}^{-\lambda t}$, $t \geq 0$. Prove that linear combinations of functions $e_\lambda, \lambda \geq 0$ form a dense set in $C[0,\infty]$. **Hint:** Treat $[0,\infty]$ as a compact topological space.

Exercise 7.6. Use Exercise 7.5 to argue as in the Hausdorff moment problem that for $x \in C[0,\infty]$, condition

$$\int_0^\infty \mathrm{e}^{-\lambda t} x(t)\,\mathrm{d}t = 0, \qquad \lambda \geq 0,$$

holds if and only if $x = 0$.

Exercise 7.7. Use the Weierstrass theorem to show the following result known as the Riemann lemma: for any $x \in C[a,b]$,

$$\lim_{n\to\infty}\int_a^b x(s)\sin ns\,\mathrm{d}s = \lim_{n\to\infty}\int_a^b x(s)\cos ns\,\mathrm{d}s = 0.$$

Hint: Prove first that the theorem holds for $x_k(s) = s^k, k \geq 0$. This can be done inductively.

Exercise 7.8. ▲ Let $C_0(0,1]$ be the space of real continuous functions on the unit interval that vanish at 0. Prove that polynomials that vanish at 0 form a linearly dense subset of this space. **Hint:** See [14] pp. 146–147.

☞ CHAPTER SUMMARY

The Weierstrass theorem says that any continuous function on a finite closed interval can be uniformly approximated, with any required accuracy, by polynomials. The Stone–Weierstrass theorem extends this result to an abstract setting, where the interval is replaced by a compact topological space, and the role of polynomials is played by a class of functions that enjoy certain properties mimicking those of polynomials. There are scores of various proofs of the latter result; the one presented in this little book could not fail to stress the importance of the completeness of the space of continuous functions.

8

Norms Do Differ

8.1 Completing Normed Spaces

What can we do if the normed space we work with is holey, and we would like to enjoy the blessings of a complete space? We simply follow the example of our ancestors, who completed the set of rationals with irrational numbers: we complete our space, making it a Banach space. For it turns out, to our benefit, that any normed space can be seen as a subspace (not a closed subspace) of a 'larger' Banach space. More specifically, for any normed linear space $\mathbb{X}$ there is a Banach space $\mathbb{Y}$ and its linear subspace $\mathbb{X}'$ with the following properties:

1. First of all, $\mathbb{X}$ is indistinguishable from $\mathbb{X}'$, that is, there is a linear map $L\colon \mathbb{X} \to \mathbb{Y}$ that is one-to-one and onto $\mathbb{X}'$, and preserves the norm:

$$\|Lx\|_{\mathbb{Y}} = \|x\|_{\mathbb{X}}; \tag{8.1}$$

   linearity simply means that

$$L(\alpha x_1 + \beta x_2) = \alpha(Lx_1) + \beta(Lx_2)$$

   for all $x_1, x_2 \in \mathbb{X}$ and $\alpha, \beta \in \mathbb{R}$ (we will have much to say about such maps in Chapter 12). We note that the symbol $+$ on the left-hand side here denotes addition in $\mathbb{X}$, but the same symbol on the right denotes addition in $\mathbb{Y}$; also, for example, multiplication αx_1 is made in $\mathbb{X}$, whereas $\alpha(Lx_1)$ is made in $\mathbb{Y}$. A similar remark concerns relation (8.1), and the subscripts stress this.
2. Second, $\mathbb{X}'$ is dense in $\mathbb{Y}$. This means that for all $y \in \mathbb{Y}$ there is a sequence $(y_n)_{n\geq 1}$ of elements of $\mathbb{X}'$ such that $\lim_{n\to\infty} \|y_n - y\|_{\mathbb{Y}}$.

Interestingly, the space $\mathbb{Y}$ is determined uniquely (up to an isomorphism): if there are two spaces $\mathbb{Y}$ with the properties listed above, they are indistinguishable, isomorphic.

The theorem presented above is proved in an abstract, though natural way, see e.g., [11]. First of all $\mathbb{X}$ is immersed into a larger space b_c of Cauchy sequences $(x_n)_{n\geq 1}$ of elements of $\mathbb{X}$ (note that to each $x \in \mathbb{X}$ one can assign the constant sequence in b_c), and then b_c is divided by an equivalence relation: two sequences are in this equivalence relation if they lie infinitely close to each other at infinity. The so-divided b_c is the $\mathbb{Y}$ searched for.

We will not repeat this proof/construction here, because in practice, at least in the practice of a freshman in functional analysis, $\mathbb{X}$ usually turns out to be a dense subspace of a natural Banach space. The aim of this chapter is to illustrate this idea.

8.2 Two Completions of the Space c_{00}

Let c_{00} be the space of sequences $x = (\xi_i)_{i\geq 1}$ such that $\xi_i = 0$ for all $i \geq i_0(x)$. A number of norms can be introduced in this space. For example, one can consider

$$\|x\|_\infty := \max_{i\geq 1} |\xi_i| \qquad \text{or} \qquad \|x\|_1 := \sum_{i=1}^{\infty} |\xi_i|$$

(the last sum has a finite number of non-zero summands). As we know from the previous section, c_{00}, which is not complete, can be completed to a Banach space. We want to stress, however, that different norms lead to different completions – this should be clear already from the first condition presented above, that is, from condition (8.1).

We claim that the completion of c_{00} in the maximum norm is c_0. To see this we note that c_{00} is a subset of c_0, and for L of point 1 of Section 8.1 one can take a map that assigns to a sequence $x \in c_{00}$ the same sequence but seen as a member of c_0. This map is obviously linear, and condition (8.1) is satisfied because the maximum norm in c_{00} is a restriction of the norm of c_0 to c_{00}. We are thus left with showing that each $x = (\xi_i)_{i\geq 1} \in c_0$ can be approximated by a sequence $(x_n)_{n\geq 1}$ of elements of c_{00}. To this end, we think of

$$x_n := (\xi_1, \xi_2, \dots, \xi_n, 0, 0, 0, \dots). \tag{8.2}$$

The so-defined x_n obviously belongs to c_{00} for each n. Moreover, the first n coordinates of $x - x_n$ are zero, and the remaining coordinates coincide with those of x. It follows that

$$\|x - x_n\|_\infty = \sup_{i\geq n+1} |\xi_i|.$$

Since x is a member of c_0 the supremum on the right-hand side converges to 0, as $n \to \infty$. This completes the proof of our first claim.

Now we claim that the completion of c_{00} in the norm $\|\cdot\|_1$ is ℓ^1. The proof is analogous. The set c_{00}, as equipped with this norm, can be seen as a subspace of ℓ^1, and thus the role of L can be played by an identity operator mapping an $x \in c_{00}$ to the same x but seen as a member of ℓ^1. Moreover, if x belongs to ℓ^1, then each x_n defined by (8.2) belongs to c_{00} and, arguing as before, we obtain

$$\|x - x_n\| = \sum_{i=n+1}^{\infty} |\xi_i|.$$

The sum featuring here converges to 0, as $n \to \infty$, because ℓ^1 is the space of absolutely summable sequences.

We have made our point: two different norms lead to two different completions of a single space.

8.3 Three Completions of the Space of Polynomials

Here is a slightly more advanced, and perhaps somewhat more interesting, example. As we have seen in Chapter 7 (see also Section 14.1, further down), each continuous function on a compact unit interval $[0, 1]$ can be approximated with any prescribed accuracy by polynomials (in the supremum norm). In other words, the completion of the linear space of polynomials, as equipped with the supremum norm (see Section 4.4.2), is $C[0, 1]$. The reader has probably noticed that our statement remains true if 'completion' is replaced by 'closure.' This shows how natural the procedure we are encountering here is. The difficulty lies in guessing the space $\mathbb{Y}$ in which we should immerse the holey normed space $\mathbb{X}$. If we guess correctly, $\mathbb{Y}$ will be the closure of $\mathbb{X}$. For example, the first condition of Section 8.1 will also be satisfied if we immerse c_{00} in the space c of sequences that have finite limits. But the second condition will fail: c is too large to be the completion of c_{00}.

Let's come back to the space of polynomials. If we equip this space with the norm defined as

$$\|x\| = \int_0^1 |x(s)|\, \mathrm{d}s,$$

its completion will not turn out to be $C[0, 1]$ but the space of functions that are absolutely integrable with respect to the Lebesgue measure on $[0, 1]$.

Since we decided to keep this book as elementary as possible, we will not discuss the details of the last statement. Instead, in the next section, we will illustrate the fact that different norms lead to different completions by examining the so-called tensor norms in the simple case of sequence spaces.

Before doing that, however, we note that a norm of a polynomial

$$x(t) = a_0 + a_1 t + \cdots + a_n t^n$$

can also be defined as follows:

$$\|x\| := \max_{i=0,1,\ldots,n} |a_i|;$$

this norm is as good as any other. A moment of reflection persuades us that, equipped with this norm, the space of polynomials is quite the same as c_{00}. Hence, the completion of this space in this norm is nothing other than c_0.

See Exercises 8.8 and 8.9 for yet other completions of c_{00}.

8.4 Tensor Products and Tensor Norms

8.4.1 Tensor Product

For $x = (\xi_i)_{i\geq 1} \in c_0$ and $y = \left(\eta_j\right)_{j\geq 1} \in \ell^1$, we define their tensor product as

$$x \otimes y = \left(\xi_i \eta_j\right)_{i,j\geq 1}. \tag{8.3}$$

Thus, the tensor product of two sequences is a matrix. For example, if both x and y are members of ℓ^1 (note that ℓ^1 can be seen as a subset of c_0) such that all their coordinates are non-negative and sum to 1, then they can be regarded as distributions of integer-valued random variables, say X and Y, so that

$$\mathbb{P}(X = i) = \xi_i, \qquad \mathbb{P}(Y = j) = \eta_j.$$

Then the matrix $x \otimes y$ is the joint distribution of X and Y as long as these random variables are independent:

$$\xi_i \eta_j = \mathbb{P}(X = i, Y = j).$$

Matrices of the form (8.3) are said to be elementary tensors, and their linear combination, that is, matrices of the form

$$m = \sum_{k=1}^{\ell} x_k \otimes y_k, \tag{8.4}$$

where $x_k \in c_0$ and $y_k \in \ell^1$ are arbitrary vectors, are termed tensors; the upper index ℓ is a natural number that depends on the tensor. The linear space of all tensors is termed the **tensor product** (of spaces c_0 and ℓ^1) and denoted

$$c_0 \otimes \ell^1.$$

It will be convenient to use the following terminology: all elements of a matrix $(m_{i,j})_{i,j\geq 1}$ with fixed index i form the matrix's ith row, and all its elements with fixed index j will be called its jth column.

Let

$$e_k := \left(\delta_{k,n}\right)_{n\geq 1} \in c_0 \cap \ell^1,$$

where $\delta_{k,n} = 1$ for $n = k$ and is zero for other n. The tensors built from $e_k, k \geq 1$ as follows,

$$e_{i,j} := e_i \otimes e_j, \qquad i, j \geq 1,$$

deserve special attention (could you interpret them probabilistically?). We note also that the representation (8.4) is not unique. For example,

$e_{2,2} = e_2 \otimes e_2$, but also $e_{2,2} = (e_1+e_2)\otimes(e_1+e_2) - e_1\otimes e_2 - e_2\otimes e_1 - e_1\otimes e_1.$

8.4.2 The Projective Tensor Norm

Is there a natural norm in the space $c_0 \otimes \ell^1$? Well, it depends on what we mean by *natural*. In what follows we will consider only norms that satisfy the following condition:

$$\|x \otimes y\| = \|x\|_{c_0} \|y\|_{\ell^1}. \tag{8.5}$$

Because of the triangle inequality, for each representation (8.4) we must have

$$\|m\| \leq \sum_{k=1}^{\ell} \|x_i\|_{c_0} \|y_i\|_{\ell^1}$$

so that the 'largest' possible candidate for a norm is

$$\|m\|_\pi = \inf \sum_{k=1}^{\ell} \|x_i\|_{c_0} \|y_i\|_{\ell^1}; \tag{8.6}$$

in this formula, the infimum is over all possible representations (8.4). A straightforward, if lengthy, calculation shows that $\|\cdot\|_\pi$ can also be written as follows:

$$\|m\|_\pi = \sum_{k=1}^{\infty} \sup_{i\geq 1} |m_{i,k}|. \tag{8.7}$$

$\|\cdot\|_\pi$ is known as the *projective norm*.

From here on, we will treat (8.7) as a definition. It is easy to check that $\|\cdot\|_\pi$ is indeed a norm. It turns out, however, that $c_0 \otimes \ell^1$ with the so-defined norm is not complete. Our aim is to show that its completion is the space of matrices

$m = (m_{i,j})_{i,j\geq 1}$ with the following properties: first, each column of m is an element of c_0, and, second,

$$\sum_{j=1}^{\infty} \sup_{i\geq 1} |m_{i,j}| < \infty. \tag{8.8}$$

The latter space will be denoted

$$c_0 \widehat{\otimes}_\pi \ell^1;$$

the norm in this space is, of course, given by

$$\|m\|_\pi := \sum_{j=1}^{\infty} \sup_{i\geq 1} |m_{i,j}|.$$

Alternatively, we could write

$$\|m\|_\pi := \sum_{j=1}^{\infty} \|m_j\|_{c_0}, \tag{8.9}$$

where m_j denotes the jth column of m. Readers should convince themselves that for elementary tensors $m = x \otimes y$ we have $\|m\|_\pi = \|x\|_{c_0}\|y\|_{\ell^1}$ and in particular that $c_0 \otimes \ell^1$ is a subset of $c_0 \widehat{\otimes}_\pi \ell^1$. Also, the reader should check that $c_0 \widehat{\otimes}_\pi \ell^1$ is a Banach space.

To show that $c_0 \widehat{\otimes}_\pi \ell^1$ indeed is a completion of $c_0 \otimes \ell^1$, we fix $m \in c_0 \widehat{\otimes}_\pi \ell^1$. For any $\epsilon > 0$ a $k \in \mathbb{N}$ can be chosen so that

$$\sum_{j=k+1}^{\infty} \sup_{i\geq 1} |m_{i,j}| < \frac{\epsilon}{2}.$$

Furthermore, an ℓ can be chosen so that for all $j \in \{1,\ldots,k\}$ we have $\sup_{i>\ell} |m_{i,j}| \leq \frac{\epsilon}{2k}$. Let $m_\epsilon = (m_{\epsilon,i,j})_{i,j\geq 1}$ be defined as follows:

$$m_{\epsilon,i,j} = \begin{cases} m_{i,j}, & j \leq k \text{ and } i \leq \ell, \\ 0, & \text{otherwise.} \end{cases} \tag{8.10}$$

In other words, $m_\epsilon = \sum_{i=1}^{\ell}\sum_{j=1}^{k} m_{i,j} e_i \otimes e_j$, and m_ϵ is a member of $c_0 \otimes \ell^1$, see Figure 8.1. Thus,

$$\begin{aligned}\|m - m_\epsilon\|_\pi &= \sum_{j=1}^{\infty} \sup_{i\geq 1} |m_{i,j} - m_{\epsilon,i,j}| \\ &= \sum_{j=1}^{k} \sup_{i>l} |m_{i,j}| + \sum_{j=k+1}^{\infty} \sup_{i\geq 1} |m_{i,j}| \leq \sum_{j=1}^{k} \frac{\epsilon}{2k} + \frac{\epsilon}{2} = \epsilon.\end{aligned}$$

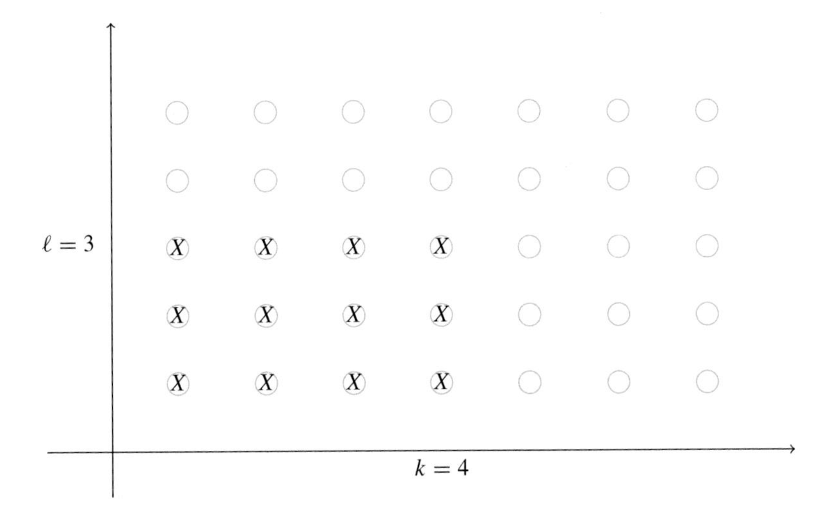

Figure 8.1 In the places marked with X, m_ϵ coincides with m, and in the remaining places it is zero. The sum of supremums of absolute values of elements of all the columns starting from (and including) the $(k+1)$st column is small (here $k = 4$). In all the first k columns the supremums of the absolute values in the places that are not marked with X are also small.

This completes the proof.

8.4.3 The Injective Tensor Norm

Does changing the order in which $\sum$ and sup appear in (8.7) lead to an interesting norm? Does the formula

$$\|m\|_\varepsilon := \sup_{i\geq 1} \sum_{j=1}^{\infty} |m_{i,j}| \tag{8.11}$$

define a norm in the first place? Fortunately, the answer is in the affirmative. In this section we will describe completion of $c_0 \otimes \ell^1$ when equipped with this norm; $\|\cdot\|_\varepsilon$ is termed the *injective norm.*

We will show that the completion of $c_0 \otimes \ell^1$ in this case is the space of matrices $m = \left(m_{i,j}\right)_{i,j\geq 1}$ such that $\sum_{j=1}^{\infty} |m_{i,j}|$ is finite for all i and the vector $Im = \left(\sum_{j=1}^{\infty} |m_{i,j}|\right)_{i\geq 1}$ belongs to c_0. This space is denoted

$$c_0 \mathbin{\widehat{\otimes}_\varepsilon} \ell^1.$$

First of all, see Exercise 8.6, $\|\cdot\|_\varepsilon$ is a norm in $c_0 \otimes \ell^1$. Also, see Exercise 8.7, $c_0 \widehat{\otimes}_\varepsilon \ell^1$ with injective norm is a Banach space. We will show that $c_0 \otimes \ell^1$ is a dense set in this space by arguing similarly as in the previous section. Namely, given $m \in c_0 \widehat{\otimes}_\varepsilon \ell^1$ and $\epsilon > 0$ we first choose an ℓ so that $\sup_{i\geq \ell+1} \sum_{j=1}^{\infty} |m_{i,j}| < \epsilon$. Then we choose a k so that $\sup_{1\leq i\leq \ell} \sum_{j=k+1}^{\infty} |m_{i,j}| < \epsilon$. Then, for m_ϵ defined by (8.10) we have

$$\begin{aligned}\|m - m_\epsilon\|_\varepsilon &= \sup_{i\geq 1} \sum_{j=1}^{\infty} |m_{i,j} - m_{\epsilon,i,j}| \\ &= \max\left(\sup_{1\leq i\leq \ell} \sum_{j=1}^{\infty} |m_{i,j} - m_{\epsilon,i,j}|,\ \sup_{i\geq \ell+1} \sum_{j=1}^{\infty} |m_{i,j} - m_{\epsilon,i,j}|\right) \\ &= \max\left(\sup_{1\leq i\leq \ell} \sum_{j=k+1}^{\infty} |m_{i,j}|,\ \sup_{i\geq \ell+1} \sum_{j=1}^{\infty} |m_{i,j}|\right) < \epsilon.\end{aligned}$$

This proves our claim.

8.4.4 Comparison of the Two Tensor Norms

Let's summarize our discussion. In the space of tensors one can introduce (at least) two norms: the projective norm and the injective norm. By completing $c_0 \otimes \ell^1$ with respect to these norms we obtain two Banach spaces: $c_0 \widehat{\otimes}_\pi \ell^1$ and $c_0 \widehat{\otimes}_\varepsilon \ell^1$.

How are these spaces related? First of all, we note the following inequality connecting the two discussed norms:

$$\|m\|_\varepsilon \leq \|m\|_\pi. \tag{8.12}$$

For its proof, we observe that $\sum_{j=1}^{\infty} |m_{i,j}| \leq \sum_{j=1}^{\infty} \sup_{i\geq 1} |m_{i,j}|$, for all i. Then, taking the supremum of the left-hand side we obtain our claim. Because of this inequality, any sequence of elements of $c_0 \otimes \ell^1$ that satisfies the Cauchy condition with respect to the projective norm satisfies the Cauchy condition for the injective norm as well. It follows that the completion of $c_0 \otimes \ell^1$ in the injective norm contains the completion of $c_0 \otimes \ell^1$ in the projective norm:

$$c_0 \widehat{\otimes}_\varepsilon \ell^1 \supset c_0 \widehat{\otimes}_\pi \ell^1.$$

Is the opposite inclusion also true? This time we need to answer in the negative. For example, consider the matrix m which on the main diagonal has the sequence $(1, \frac{1}{2}, \frac{1}{3}, \ldots)$ and all its remaining entries are zero. Then, m belongs to $c_0 \widehat{\otimes}_\varepsilon \ell^1$, because Im coincides with $(1, \frac{1}{2}, \frac{1}{3}, \ldots)$. On the other hand, m does not belong to $c_0 \widehat{\otimes}_\pi \ell^1$ because $\sum_{j=1}^{\infty} \sup_{i\geq 1} |m_{i,j}| = \sum_{j=1}^{\infty} \frac{1}{j} = \infty$. Thus, the situation is quite analogous to that described in Sections 8.2 and 4.4.2: the stronger the norm, the smaller is completion.

The main take-home message from this chapter, however, is that

> completion of a linear space depends crucially on the choice of the norm involved.

A systematic exposition of the theory of tensor products of Banach spaces and of tensor norms can be found in the book [37] – the lion's share of this theory is due to A. Grothendieck (1928–2014), one of the giants of twentieth-century mathematics.

8.5 Exercises

Exercise 8.1. To a vector $x = (\xi_0, \xi_1, \xi_2, \dots, \xi_n, 0, \dots) \in c_{00}$ assign

$$\|x\| = \max_{t\in[0,1]} \left| \sum_{i=0}^{n} \xi_i t^i \right|.$$

Check to see that this is a norm, and find the completion of c_{00} in this norm.

Exercise 8.2. Repeat the analysis of Section 8.4.4 for the norms of Section 8.2 to explain why c_0 (as a set) contains ℓ^1.

Exercise 8.3. Elements of $c_0 \widehat{\otimes}_\pi \ell^1$ can be thought of as sequences $(x_n)_{n\geq 1}$ of elements of c_0 such that $\sum_{n=1}^{\infty} \|x_n\|_{c_0} < \infty$, that is, as absolutely summable sequences with values in c_0. Provide a similar description for the elements of $c_0 \widehat{\otimes}_\varepsilon \ell^1$.

Exercise 8.4. Check to see that $\|\cdot\|_\pi$ defined by (8.7) is a norm and satisfies condition (8.5).

Exercise 8.5. Show that the space of matrices satisfying (8.8) is a Banach space.

Exercise 8.6. Check that $\|\cdot\|_\varepsilon$ is a norm in $c_0 \otimes \ell^1$ and satisfies condition (8.5).

Exercise 8.7. Show that $c_0 \widehat{\otimes}_\varepsilon \ell^1$ with injective norm is a Banach space.

Exercise 8.8. Let c_{00} be equipped with

$$\|x\| = \|(\xi_i)_{i\geq 1}\| = \sum_{i=1}^{\infty} 2^n |\xi_i|$$

(this sum is finite because it involves a finite number of summands). Provide a characterization of the completion of c_{00} with this norm.

Exercise 8.9. Repeat the previous exercise for the following norm:

$$\|x\| = \|(\xi_i)_{i\geq 1}\| = \sum_{i=1}^{\infty} |\xi_{2i}| + \sup_{i\geq 0} |\xi_{2i+1}|.$$

Exercise 8.10. Let $C[0,1]$ be the space of continuous functions on the unit interval, and let $C[0,1] \otimes C[0,1]$ be the space of functions z on $[0,1] \times [0,1]$ that are of the form $z(s,t) = \sum_{i=1}^{n} x_i(s)y_i(t), s,t \in [0,1]$, where the natural number n and functions $x_i, y_i \in C[0,1]$ can be chosen arbitrarily. Functions of the particular form $z(s,t) = x(s)y(s), s,t \in [0,1]$ and $x, y \in C[0,1]$ are called simple tensors and denoted $x \otimes y$, and $C[0,1] \otimes C[0,1]$ is known as the tensor product of two copies of $C[0,1]$. Check to see that $\|z(t,s)\| := \max_{s,t\in[0,1]} |z(s,t)|$ defines a tensor norm in that $\|x \otimes y\| = \|x\|_{C[0,1]} \|y\|_{C[0,1]}$. Prove that the completion of $C[0,1] \otimes C[0,1]$ (in this tensor norm) is the space of continuous functions on $[0,1] \times [0,1]$. **Hint:** Convergence in this norm is just the uniform convergence, hence limits of Cauchy sequences have to be continuous functions. For the converse, use the Stone–Weierstrass theorem.

☞ CHAPTER SUMMARY

A normed linear space can be (uniquely) completed to a Banach space. However, whereas a Banach space is a match for its practically unique norm (see also Section 15.8), there are many possible norms that can be used in a linear space, and depending on the choice of norm we obtain many different completions of a single space. This phenomenon is discussed first in the case of c_{00} and in the case of the space of polynomials. The injective and projective tensor norms, which show up naturally in the tensor product $c_0 \otimes \ell^1$, illustrate this principle further, but they have their own importance, reaching far beyond the scope of this book.

9
Hilbert Spaces

A Hilbert space is a complete linear space with the norm constructed from a scalar product. Hence, it is a very particular Banach space, because the norm that comes from a scalar product very much resembles the norm of Euclidean spaces with which we are so familiar (see Figure 9.1).

9.1 Unitary Spaces

9.1.1 Definition and Basic Properties

A unitary space is a linear space $\mathbb{X}$, equipped with a scalar product, that is, with a function $\mathbb{X} \times \mathbb{X} \to \mathbb{R}$ (in this book we restrict ourselves to linear spaces over the space of reals, and thus our scalar product also has real values) such that:

(s1). $(x + y, z) = (x, z) + (y, z)$,
(s2). $(\alpha x, y) = \alpha(x, y)$,
(s3). $(x, x) \geq 0$,
(s4). $(x, x) = 0$ if and only if $x = 0$,
(s5). $(x, y) = (y, x)$;

here, x, y and z are arbitrary elements of $\mathbb{X}$, α is an arbitrary scalar, and (x, y) is the value of the scalar product of vectors x and y.

The Euclidean space $\mathbb{R}^k$ is an example of a unitary space, with scalar product

$$(x, y) := \sum_{i=1}^{k} \xi_i \eta_i; \tag{9.1}$$

checking conditions (s1)–(s5) is particularly simple in this case.

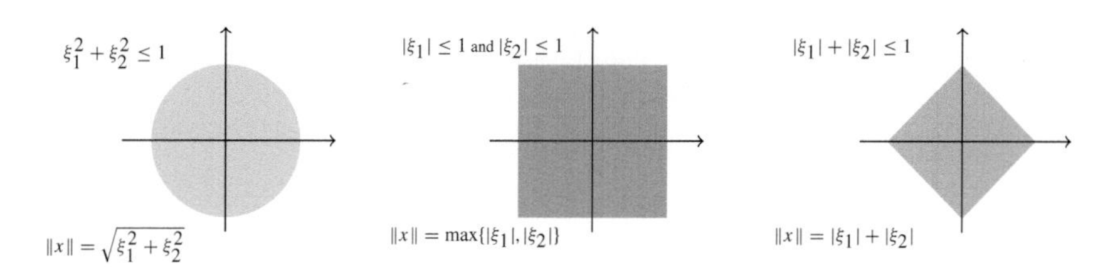

Figure 9.1 (Closed) unit balls in $\mathbb{R}^2$ related to various norms: (a) the Euclidean norm (coming from the scalar product), (b) the supremum norm, and (c) the ℓ^1-type norm.

9.1 Lemma (Cauchy–Schwarz) *For all x and y in a unitary space, we have*

$$|(x,y)| \le \sqrt{(x,x)}\sqrt{(y,y)}.$$

Proof Fix x and y, and think of the function

$$\mathbb{R} \ni t \mapsto (tx+y, tx+y) \in \mathbb{R}.$$

Because of (s3), this function has non-negative values. Moreover, simple calculations based on (s1), (s2) and (s5) show that

$$(tx+y, tx+y) = t^2(x,x) + 2t(x,y) + (y,y).$$

We can exclude the case where $(x,x) = 0$, because then the Cauchy–Schwarz inequality is obvious. Also, if $(x,x) > 0$ (condition (s3) guarantees that $(x,x) < 0$ is impossible), the graph of this quadratic function is a parabola with arms directed upwards. Since we know that this parabola lies above the t-axis, its discriminant is at most zero, that is,

$$[(x,y)]^2 - (x,x)(y,y) \le 0.$$

This completes the proof. □

For example, in the case of $\mathbb{R}^k$, the Cauchy–Schwarz inequality takes the form

$$\left|\sum_{i=1}^{k} \xi_i \eta_i\right| \le \sqrt{\sum_{i=1}^{k} \xi_i^2}\sqrt{\sum_{i=1}^{k} \eta_i^2}. \tag{9.2}$$

This inequality in turn allows introducing another, more interesting unitary space. This is the space ℓ^2 of square integrable sequences, that is, the sequences $x = (\xi_i)_{i\ge 1}$ such that

$$\sum_{i=1}^{\infty} \xi_i^2 < \infty.$$

This space is equipped with a natural scalar product, defined as follows:

$$(x, y) := \sum_{i=1}^{\infty} \xi_i \eta_i. \tag{9.3}$$

How do we know that the series on the right converges? This can be deduced from (9.2). Indeed, for any $x, y \in \ell^2$, we have

$$\sum_{i=1}^{k} |\xi_i|\,|\eta_i| \le \sqrt{\sum_{i=1}^{k} \xi_i^2}\sqrt{\sum_{i=1}^{k} \eta_i^2} \le \sqrt{\sum_{i=1}^{\infty} \xi_i^2}\sqrt{\sum_{i=1}^{\infty} \eta_i^2} < \infty.$$

Hence, k being arbitrary, we conclude that

$$\sum_{i=1}^{\infty} |\xi_i|\,|\eta_i| \le \sqrt{\sum_{i=1}^{\infty} \xi_i^2}\sqrt{\sum_{i=1}^{\infty} \eta_i^2} < \infty,$$

showing that our series converges, and converges absolutely. With this inequality under our belt, we easily become experts in unitary spaces and are able to check that (9.3) indeed defines a scalar product.

As an immediate consequence of the Cauchy–Schwarz inequality, we conclude that unitary spaces are automatically normed spaces. Indeed, the formula

$$\|x\| := \sqrt{(x, x)} \tag{9.4}$$

defines a natural norm in a unitary space. In the case of ℓ^2, for instance, we obtain thus the norm

$$\|(\xi_i)_{i \ge 1}\| := \sqrt{\sum_{i=1}^{\infty} \xi_i^2}, \tag{9.5}$$

which should look appealing.

Below, we check that the function defined in (9.4) satisfies the triangle inequality; the reader should check the remaining properties of a norm. In terms of this scalar-product-induced norm, the Cauchy–Schwarz inequality reads

$$|(x, y)| \le \|x\|\,\|y\|, \qquad x, y \in \mathbb{X}.$$

Thus, for any x and y in $\mathbb{X}$,

$$\begin{aligned}
\|x + y\|^2 &= (x + y, x + y) = (x, x) + 2(x, y) + (y, y) \\
&\le (x, x) + 2\|x\|\,\|y\| + (y, y) \\
&= \|x\|^2 + 2\|x\|\,\|y\| + \|y\|^2 \\
&= (\|x\| + \|y\|)^2.
\end{aligned}$$

This, however, means that $\|x + y\| \le \|x\| + \|y\|$.

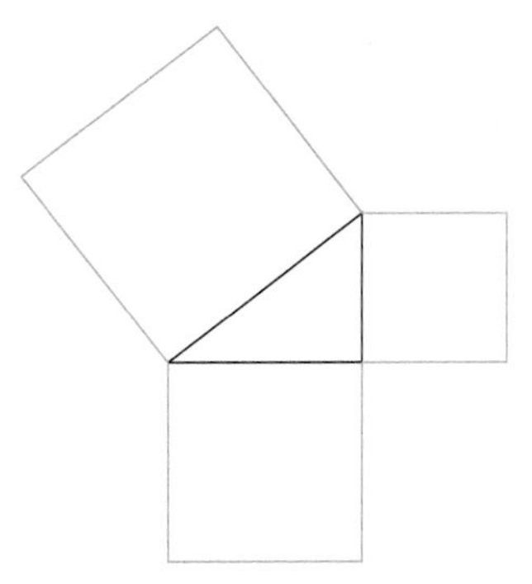

Figure 9.2 Sum of the areas of smaller squares is the area of the larger square because . . .

9.1.2 Simple Applications of the Scalar Product

Some students have seen the definition of a scalar product so many times that they start to treat this definition as obvious. We should not forget, though, that we have waited for its formulation for a good couple of thousand years – since the time our ancestors started to be interested in geometry. Yes, I do mean geometry. Functional analysis is often treated as a generalization of the ancient art of earth measuring. I think we will be able to appreciate the value of this notion more when we are presented with a scalar-product based proof of the theorem usually attributed to Pythagoras (but known at least one thousand years before him[3]): in the right triangle, the area of the square whose side is the hypotenuse is equal to the sum of the areas of the squares on the other two sides. There are apparently dozens of dozens of proofs of this result; Figures 9.2–9.4 present the proof that is due to Euclid himself. Notably, to come up with some of these required large amounts of perceptiveness and ingenuity. Meanwhile, if you are familiar with scalar products, you can prove this theorem even if you are not that perceptive or ingenious. For, in a unitary space, of which $\mathbb{R}^2$ with the scalar product (9.1) is an example, Pythagoras's theorem says simply that whenever vectors x and y are perpendicular (think of them as the sides of the triangle next the the right angle), that is, whenever $(x, y) = 0$, we have

$$\|x + y\|^2 = \|x\|^2 + \|y\|^2;$$

[3] Nevertheless, you should not believe some Russian jokers who claim that the theorem was first proved by a Pyetya Goras, a famous Russian mathematician.

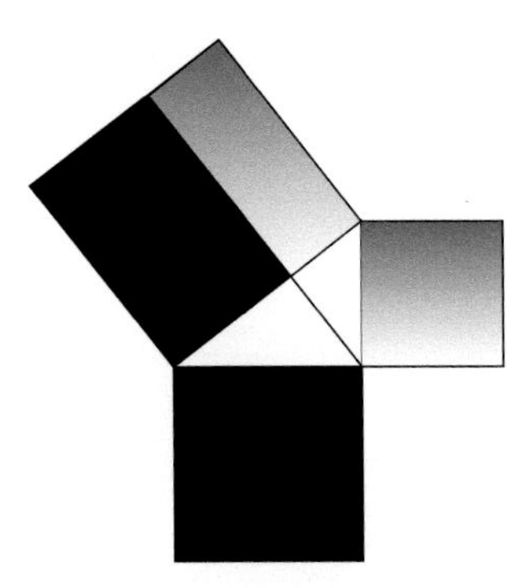

Figure 9.3 ... like-colored rectangles have equal areas; for example, gray-shaded areas are the same because ...

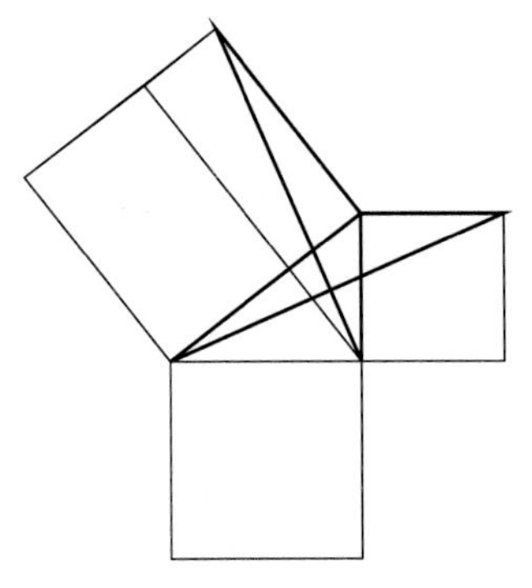

Figure 9.4 ... triangles marked with thick lines are congruent (one can be obtained from the other by rotating it around the common vertex), and the area of either of them is half of the area of the corresponding rectangle.

$x + y$, obviously, plays the role of the hypotenuse here. The proof is really just a straightforward calculation:

$$x \perp y \implies \|x + y\|^2 = (x + y, x + y) = (x, x) + \underbrace{2(x, y)}_{=0} + (y, y)$$
$$= \|x\|^2 + \|y\|^2.$$

Encouraged by this simplicity, let's prove another result, known as the parallelogram law. This law says that the sum of the squares of the lengths of the four sides of a parallelogram equals the sum of the squares of the lengths of the two diagonals. A standard proof of this result consists of applying Pythagoras's theorem or the law of cosines. However, let's show it directly and painlessly (i.e., without bothering to think). We claim namely that in a unitary space $\mathbb{X}$ the following formula is true:

$$\|x + y\|^2 + \|x - y\|^2 = 2\big[\|x\|^2 + \|y\|^2\big] \tag{9.6}$$

for all $x, y \in \mathbb{X}$, where $\|\cdot\|$ is the norm introduced by means of the scalar product involved. Again, the proof is nothing but a calculation, for we have

$$\|x + y\|^2 = (x + y, x + y) = (x, x) + 2(x, y) + (y, y),$$
$$\|x - y\|^2 = (x - y, x - y) = (x, x) - 2(x, y) + (y, y),$$

and the claim is obtained by adding both sides of these equations.

Some may argue that old, geometric, proofs are more beautiful and more noteworthy. Indeed, there is some merit in this statement. But we should not forget this: since the machinery of scalar products allows us to prove known theorems with such ease, what profound theorems might we prove with some effort!

9.2 Distance Minimizing Elements

As the reader probably perceives, unitary spaces that are not complete will not be much appreciated in this book. Rather, we quickly proceed to take a closer look at spaces that are not holey.

Here is a crucial definition.

9.2 Definition Let $\mathbb{H}$ be a unitary space and let $\|\cdot\|$ be the unitary norm defined by (9.4). If $\mathbb{H}$ with this norm is a Banach space, then $\mathbb{H}$ is said to be a Hilbert space.

It is perhaps worth stressing that for a unitary space to be a Hilbert space it does not suffice for it to be complete with respect to just any norm: it must be complete with respect to the norm that is related to the scalar product by (9.4).

We note also that, in saying that an $\mathbb{H}$ is a Hilbert space, it would be more proper to specify the scalar product or the unitary norm that makes $\mathbb{H}$ complete. In fact, it would be even more proper to say that a pair, a linear space plus the scalar product, forms a Hilbert space. In practice, however, the definition of the scalar product involved is often clear from the context, and a statement like 'ℓ^2 is a Hilbert space' usually does not lead to misunderstandings. The situation is in fact quite similar to that with Banach spaces: in saying that, for instance, c_0 is a Banach space, we more often than not have the usual supremum norm in mind. Hilbert spaces are as inseparably linked to their scalar products as Banach spaces are to their norms.

The space ℓ^2 of Section 9.1.1 is a basic and typical example of a Hilbert space. We omit the proof of the fact that this space is complete (when equipped with the norm (9.5) which comes from the scalar product (9.3)), because it is analogous to that of completeness of ℓ^1.

The space $L^2(S, \mathcal{F}, \mu)$ of (equivalence classes of) square integrable functions on a measure space $(S, \mathcal{F}, \mu)$ is even more important. The proof that this space is complete resembles the argument presented in Section 4.3.5, and is hence omitted here. We stress merely once again that, as is the case with the space of type L^1, strictly speaking, $L^2(S, \mathcal{F}, \mu)$ is not a space of square integrable functions but of their equivalence classes: two functions are thought of as equivalent if they differ on a set of μ measure zero. We also note that the scalar product in $L^2(S, \mathcal{F}, \mu)$ is defined by

$$(x, y) = \int_S xy \, \mathrm{d}\mu.$$

When $\mathcal{F}$ and μ are clear from the context, which – in this book – is the case when S is a subset of some $\mathbb{R}^k$ and μ is the Lebesgue measure, as restricted to this subset, we simply write $L^2(S)$. In the case of intervals $S = [a, b]$, we write $L^2[a, b]$ rather than $L^2([a, b])$.

As we have seen in the case of Pythagoras's theorem and the parallelogram law, the geometry of Hilbert spaces is familiar to us. Here is another example confirming this rule.

9.3 Theorem *Let C be a non-empty, closed and convex subset of a Hilbert space $\mathbb{H}$, and let $x \notin C$. There exists precisely one element $y \in C$ with the following property:*

$$\|x - y\| = d := \inf_{z \in C} \|x - z\|.$$

In other words, y is a unique point of C that minimizes the distance to x. To recall, a set C is said to be convex if together with any two elements x and y, it contains the entire segment with ends at x and y, that is, it contains all the vectors of the form $z = \alpha x + (1 - \alpha)y, \alpha \in [0, 1]$.

Proof (See Figure 9.5.) Because of the parallelogram law, for any two vectors z and z' in C we have

$$\begin{aligned}
\|z - z'\|^2 &= \|(z - x) + (x - z')\|^2 \\
&= 2\{\|z - x\|^2 + \|z' - x\|^2\} - \|z + z' - 2x\|^2 \\
&= 2\{\|z - x\|^2 + \|z' - x\|^2\} - 4\left\|\frac{z + z'}{2} - x\right\|^2 \\
&\le 2\{\|z - x\|^2 + \|z' - x\|^2\} - 4d^2,
\end{aligned} \tag{9.7}$$

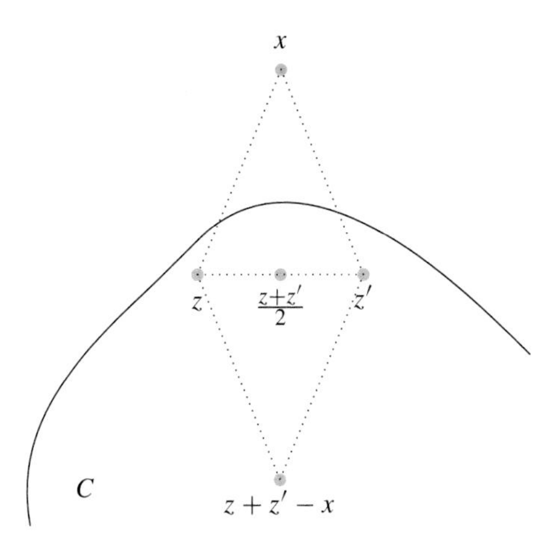

Figure 9.5 If z and z' belong to C, then so does $\frac{z+z'}{2}$. Then the distance from x to $z+z'-x$, being twice the distance between x and $\frac{z+z'}{2}$, is at least $2d$. Therefore, by the parallelogram law, if the distance between x and z and that between x and z' are close to d, then the distance between z and z' must be nearly zero.

because $\frac{z+z'}{2}$ belongs to C.

By the definition of d we know that there is a sequence $(z_n)_{n\geq 1}$ of the elements of C such that $\lim_{n\to\infty} \|z_n - x\| = d$. We will show that $(z_n)_{n\geq 1}$ is a Cauchy sequence. To this end, given $\epsilon > 0$ we choose n_0 so that for $n \geq n_0$, the squared distance $\|z_n - x\|^2$ is smaller than $d^2 + \frac{\epsilon}{4}$. Using (9.7) with $z := z_n$ and $z' := z_m$ we see that $\|z_n - z_m\|^2 \leq \epsilon$ for $n, m \geq n_0$ – and this was our goal.

Since $\mathbb{H}$ is a Hilbert space, the sequence $(z_n)_{n\geq 1}$ has a limit, and this limit belongs to C, C being closed by assumption. But for $y := \lim_{n\to\infty} z_n$, we have $\|x - y\| = \lim_{n\to\infty} \|x - z_n\| = d$. This shows that y described in the statement of the theorem exists; this y is the limit of $(z_n)_{n\geq 1}$.

We still need to argue that there is only one y with this property. However, if $\|y' - x\| = d$ for some y', then (9.7) with $z = y$ and $z' = y'$ implies $\|y - y'\| = 0$, that is, $y' = y$. □

We complete this section with the following remark. Many theorems in mathematical analysis are variations on the fact that a continuous function defined on a compact set attains its extremal values. The result presented above differs from these theorems: there is no assumption that C is compact here. This is in fact its greatest value, because compact sets are 'practically finite dimensional,' whereas sets of interest in Hilbert spaces (like its subspaces, see

Section 9.3) rarely posses this property. Hence, we may summarize Theorem 9.3 by saying that in Hilbert spaces assumptions of closedness and convexity of C successfully replace that of compactness.

9.2.1 Remarks, an Example and a Couterexample

In general Banach spaces that are not Hilbert spaces, Theorem 9.3 is no longer true. For example, the non-negative sequences with entries summing to 1 form a non-empty, convex and closed set C in ℓ^1. Nevertheless, for any negative sequence x in l^1, its distance from C equals $\|x\| + 1$ and, at the same time, the distance of x from any element of C is also $\|x\| + 1$. Here, we conclude, the element that minimizes the distance is by no means unique: all elements of C minimize the distance from x. We thus see that 'the shape of the norm,' or, put another way, the geometry of Hilbert space, is a key assumption for Theorem 9.3 – for Banach spaces with different norms the theorem need not be true.

An examination of the proof shows also that the argument fails if the space is not complete – without completeness we would not have been able to conclude that $(z_n)_{n\geq 1}$ converges, and thus would be unable to find y. In this context, we look at the following example – let $\mathbb{H}$ be equal to $\mathbb{R}^2$ and let C be the closed unit ball $\{(\xi_1, \xi_2); \xi_1^2 + \xi_2^2 \leq 1\}$ without the point $(0, 1)$. This set is non-empty and convex but not closed, and because of that there is no point in C that minimizes the distance from point $(0, 2)$. In the place where the minimizing point should be there is a hole.

Let's also have a look at the set in the same space defined as follows:

$$C = \{(\xi_1, \xi_2); \xi_1 \in [-1, 1], \xi_2 = \pm\xi_1\};$$

this set is X-shaped and its center lies at the origin. It is easy to see that C is closed but not convex. Therefore, depending on which point outside of C we choose there can be one or two points that minimize the distance from C.

Finally, let's examine a simple positive example in which the distance minimizing point can be found explicitly. Let $\mathbb{H}$ be a Hilbert space (or just a unitary space) and let C be the closed unit ball, that is the set of all $y \in \mathbb{H}$ such that $\|y\| \leq 1$. Also, let $x \notin C$, that is, $\|x\| > 1$. Which element of $y \in C$ is the closest to x? Our intuition tells us that it is $y := \frac{x}{\|x\|}$ that has this property. Let's check whether we are right. For any $y \in C$, we have

$$\begin{aligned}\|x - y\|^2 &= \|x\|^2 - 2(x, y) + \|y\|^2 \geq \|x\|^2 - 2\|x\| \, \|y\| + \|y\|^2 \\ &= (\|x\| - \|y\|)^2 \geq (\|x\| - 1)^2.\end{aligned}$$

Furthermore, for $y := \frac{x}{\|x\|}$, all inequalities become equalities in this calculation. Our intuition was right.

9.3 Projection on a Subspace

The case where C of the previous section is a closed subspace of $\mathbb{H}$ is of special interest. We will write $\mathbb{H}_1$ instead of C to stress that we are dealing with this case.

9.4 Theorem *For any x in $\mathbb{H}$ there is a unique vector $Px \in \mathbb{H}_1$ such that $x - Px$ is perpendicular to all $z \in \mathbb{H}_1$, that is, for $z \in \mathbb{H}_1$ we have $(x - Px, z) = 0$. The vector Px minimizes the distance from $\mathbb{H}_1$ to x, and is referred to as the (orthogonal) projection of x onto $\mathbb{H}_1$.*

Proof For x in $\mathbb{H}_1$ we obviously take $Px = x$, and it is easy to see that there are no other choices. For, were Px not equal to x, we would consider $z := x - Px \in \mathbb{H}_1$ to see that $(z, x - Px) \neq 0$.

Hence, let's take $x \notin \mathbb{H}_1$ and let y be the element that minimizes the distance of $\mathbb{H}_1$ from x. Let $z \neq 0$ belong to $\mathbb{H}_1$. One the one hand, the function

$$f(t) := \|x - y + tz\|^2$$

of real variable t attains its minimum at $t_{\min} = 0$, because $y - tz$ belongs to $\mathbb{H}_1$, and y minimizes the distance of x from $\mathbb{H}_1$. On the other hand, doing the calculations we went through twice already we check that

$$f(t) = \|x - y\|^2 + 2t(z, x - y) + t^2\|z\|^2.$$

This shows that f is a quadratic and that its minimum is attained at $t_{\min} = -\frac{(z,x-y)}{\|z\|^2}$. It follows that $(x - y, z) = 0$, that is, $Px := y$ has the property stated in the theorem.

We still need to check that there is no other possible choice for Px. To this end, let's suppose that there is another $y' \in \mathbb{H}_1$ such that $(x - y', z) = 0$ as long as $z \in \mathbb{H}_1$. Since $y - y'$ belongs to $\mathbb{H}_1$, we have

$$\|x - y\|^2 = \|x - y'\|^2 + 2(x - y', y' - y) + \|y' - y\|^2 = \|x - y'\|^2 + \|y' - y\|^2.$$

Were $y' \neq y$ we would conclude that $\|x - y\| > \|x - y'\|$, contradicting the definition of y. There is thus no other choice for Px; the point we are looking for has to minimize the distance. □

For example, let $\mathbb{H}_1$ be the subspace of ℓ^2 of sequences $(\eta_i)_{i\geq 1}$ satisfying

$$\eta_{2i} = \eta_{2i-1}, \qquad i \geq 1. \tag{9.8}$$

(We leave it to the reader to check that this is a closed subspace of $\mathbb{H}$.) Having given $x = (\xi_i)_{i\geq 1} \notin \mathbb{H}_1$, we look for Px in $\mathbb{H}_1$. This can be done in at least two ways. First, let's think of the squared distance between x and a y in $\mathbb{H}_1$:

$$\|x-y\|^2 = \sum_{j=1}^{\infty}(\xi_j - \eta_j)^2 = \sum_{i=1}^{\infty}\Big[(\xi_{2i}-\eta_{2i})^2 + (\xi_{2i-1}-\eta_{2i})^2\Big].$$

In this calculation, we combined two neighboring coordinates together and used the defining property of $\mathbb{H}_1$. To minimize $\|x-y\|^2$ it suffices to minimize all the summands, for the η_{2i} can be chosen arbitrarily and independently of each other. In each summand, the numbers ξ_{2i} and ξ_{2i-1} are given. We are thus facing the following task: having given a and b, choose u so that

$$(a-u)^2 + (b-u)^2$$

is the smallest. This we were taught in high school: the solution is $u = \frac{a+b}{2}$. Thus, coming back to our problem we see that the projection of x on $\mathbb{H}_1$ is given by

$$Px = \tfrac{1}{2}(\xi_1+\xi_2, \xi_1+\xi_2, \xi_3+\xi_4, \xi_3+\xi_4, \xi_5+\xi_6, \dots).$$

Let's solve this problem differently now, using the characterization given in Theorem 9.4. We begin by noting that the vector

$$z_1 = (1,1,0,0,0,\dots),$$

all of whose coordinates starting from the third are zero, is a member of $\mathbb{H}_1$. The $Px = (\eta_i)_{i\geq 1}$ we are looking for must thus satisfy

$$0 = (x - Px, z_1) = (\xi_1 - \eta_1)\cdot 1 + (\xi_2 - \eta_2)\cdot 1.$$

Since $\eta_2 = \eta_1$ (because Px is to belong to $\mathbb{H}_1$), this forces

$$\eta_1 = \eta_2 = \tfrac{1}{2}(\xi_1+\xi_2).$$

The same reasoning can be applied to the vector $z_i, i \geq 2$, which has all coordinates equal to zero except for the $2i-1$st and $2i$st coordinates, which are 1 – to yield

$$\eta_{2i-1} = \eta_{2i} = \tfrac{1}{2}(\xi_{2i}+\xi_{2i-1}).$$

This of course coincides with the solution obtained with the first method.

9.4 Connection with Conditional Expectations

Here is a much more interesting example. Let X and Y be two random variables with natural values, defined on the same probability space. We assume furthermore, for simplicity of exposition, that

$$\Pr(X = i, Y = j) = p_{i,j} > 0, \qquad i, j \geq 1.$$

Finally, we assume that X and Y have second moments, that is, that

$$\sum_{i=1}^{\infty} i^2 \Pr(X = i) < \infty \qquad \text{and} \qquad \sum_{i=1}^{\infty} i^2 \Pr(Y = i) < \infty. \tag{9.9}$$

The space $\mathbb{H}$ of matrices $x = \left(\xi_{i,j}\right)_{i,j\geq 1}$ for which the sum

$$\sum_{i,j=1}^{\infty} \xi_{i,j}^2 p_{i,j} \tag{9.10}$$

is finite is a Hilbert space with scalar product

$$(x, y) = \sum_{i,j=1}^{\infty} \xi_{i,j} \eta_{i,j} p_{i,j}$$

(see Exercices 9.7–9.9), and the sum in (9.10) is the square of the unitary norm of the matrix x.

Random variables X and Y can be seen as matrices: X can be identified with the matrix $x = \left(\xi_{i,j}\right)_{i,j\geq 1}$ defined by $\xi_{i,j} = i$, and Y can be identified with $y = \left(\eta_{i,j}\right)_{i,j\geq 1}$, where $\eta_{i,j} = j$. Assumption (9.9) tells us now that

$$\sum_{i,j=1}^{\infty} \xi_{i,j}^2 p_{i,j} = \sum_{i=1}^{\infty} i^2 \sum_{j=1}^{\infty} p_{i,j} = \sum_{i=1}^{\infty} i^2 \Pr(X = i) < \infty,$$

which shows that x belongs to $\mathbb{H}$. Similarly, we check that $y \in \mathbb{H}$.

With these preparations out of the way, we consider the subspace $\mathbb{H}_1$ of matrices $z = \left(\zeta_{i,j}\right)_{i,j\geq 1}$ with entries that do not depend on i. In other words, in the convention of Section 8.4 any column of $z \in \mathbb{H}_1$ is composed of identical entries. It is easy to check that this is indeed a closed subspace of $\mathbb{H}$, and we obviously have $y \in \mathbb{H}_1$. Our goal is to find the projection of x on $\mathbb{H}_1$. This matrix clearly does not belong to this subspace; on the contrary, the entries of x depend solely on i.

To this end, we fix a natural k and think of the matrix $z_k = (\zeta_{i,j})_{i,j\geq 1}$ with its kth column composed of 1's and all the remaining entries equaling zero. In other words, $\zeta_{i,j} = 1$ as long as $j = k$, and zero otherwise. This is an element of $\mathbb{H}$ because $\sum_{i=1}^{\infty} \sum_{j=1}^{\infty} \zeta_{i,j} p_{i,j} = \sum_{i=1}^{\infty} p_{i,k} = P(Y = k) < \infty$; it is also an element of $\mathbb{H}_1$. Therefore, for $Px = \left(\gamma_{i,j}\right)_{i,j\geq 1}$, we should have

$$0 = (x - Px, z_k) = \sum_{i,j} (\xi_{i,j} - \gamma_{i,j}) \zeta_{i,j} p_{i,j} = \sum_{i=1}^{\infty} (\xi_{i,k} - \gamma_{i,k}) p_{i,k},$$

because $\zeta_{i,j}$ is non-zero and equal to one if and only if $j = k$. Since $\gamma_{i,k}$ should not depend on i, and $\xi_{i,k} = i$, this yields

$$\gamma_{i,k} = \gamma_k = \frac{\sum_{i=1}^{\infty} i p_{i,k}}{\sum_{i=1}^{\infty} p_{i,k}}.$$

The so-obtained quantity is known in probability theory as the expected value of the random variable X given that $Y = k$.

Isn't it beautiful that the seemingly abstract procedure of projecting vectors onto a subspace leads to an intuitionally pleasing result? The idea presented here goes much beyond this particular example, and allows us in particular to define the so-called conditional expected values of random variables with respect to other random variables (which do not necessarily have natural values) and with respect to σ-fields, see e.g., [11, 23]. This beautiful idea is apparently due to A.N. Kolmogorov (1903–1987), a famous and prolific Russian mathematician.

9.5 Exercises

Exercise 9.1. Prove that (9.4) defines a norm.

Exercise 9.2. Let $(x_n)_{n\ge 1}$ and $(y_n)_{n\ge 1}$ be two sequences of elements of a Hilbert space such that $\|x_n\| = \|y_n\| = 1, n \ge 1$ and $\lim_{n\to\infty} \|x_n + y_n\| = 2$. Check that then $\lim_{n\to\infty} \|x_n - y_n\| = 0$.

Exercise 9.3. Show that in unitary spaces the following relation, sometimes termed the polarization formula, is true:

$$(x, y) = \frac{\|x + y\|^2 - \|x - y\|^2}{4}, \qquad x, y \in \mathbb{X}.$$

Exercise 9.4. Check to see that ℓ^2 is a unitary space, that is, that equation (9.3) defines a scalar product.

Exercise 9.5. Let $\mathbb{H}$ be a unitary space, and let $(x_n)_{n\ge 1}$ and $(y_n)_{n\ge 1}$ be two sequences of elements of this space such that the limits $\lim_{n\to\infty} x_n =: x$ and $\lim_{n\to\infty} y_n =: y$ exist. Show that $\lim_{n\to\infty}(x_n, y_n) = (x, y)$.

Exercise 9.6. Let a natural number k be fixed, and suppose that $p_1, \ldots, p_k$ are positive numbers. Check to see that the equation

$$\sum_{i=1}^{k} \xi_i \eta_i p_i$$

defines a scalar product in $\mathbb{R}^k$. Describe the figure formed by the unit ball in the Hilbert space related to this product in the cases $k = 1, 2, 3$.

Exercise 9.7. Let $p = (p_i)_{i \geq 1}$ be a sequence of positive numbers. We consider the space ℓ^2_p of sequences $(\xi_i)_{i \geq 1}$ such that the sum

$$\sum_{i=1}^{\infty} \xi_i^2 p_i \tag{9.11}$$

is finite. Using the previous problem prove that the equation

$$(x, y) = \sum_{i=1}^{\infty} \xi_i \eta_i p_i$$

defines a scalar product in ℓ^2_p, and that the sum (9.11) is the squared unitary norm of $(\xi_i)_{i \geq 1}$.

Exercise 9.8. Suppose that in the set-up of the previous exercise, we additionally have $\sum_{i=1}^{\infty} p_i = 1$. Two elements of ℓ^2_p, say x and y, can then be identified with random variables, say X and Y, defined on the probability space $\Omega = \mathbb{N}$ with $\Pr(\{i\}) = p_i$. Check to see that X and Y have second moments. Prove also that

$$(x, y) = E\, XY.$$

Exercise 9.9. Prove that ℓ^2 and ℓ^2_p (see Exercise 9.7) are Hilbert spaces.

Exercise 9.10. Persuade yourself that neither the norm in ℓ^1 nor that in E of Exercise 4.13 come from a scalar product. **Hint:** Find vectors for which the parallelogram law fails.

Exercise 9.11. Let $\mathbb{H}$ be a Hilbert space, and let $\mathbb{H}_1$ be its closed subspace.

(a) Prove that the set $\mathbb{H}_1^{\perp}$ of vectors that are perpendicular to all vectors in $\mathbb{H}_1$ is a closed subspace of $\mathbb{H}$.
(b) Describe $\mathbb{H}_1^{\perp}$ for $\mathbb{H}_1 \subset \ell^2$ defined in (9.8).
(c) Let x belong to $\mathbb{H}$. Prove that Px is a projection of x on $\mathbb{H}_1$ if and only if $x - Px$ is a projection of x on $\mathbb{H}_1^{\perp}$. Use this information to find the projection on $\mathbb{H}_1^{\perp}$ of point (b).

Exercise 9.12. Let $\mathbb{H}_1$ be the closed subspace of ℓ^2 of sequences $(\eta_i)_{i \geq 1}$ such that $\eta_{3i} = \eta_{3i-1} = \eta_{3i-2}$ for all $i = 1, 2, \ldots$. Can you guess the form of projection of a vector $(\xi_i)_{i \geq 1}$ on this subspace? Prove your conjecture.

Exercise 9.13. Let $\mathbb{H}_1$ be the closed subspace of ℓ^2 of sequences $(\eta_i)_{i\geq 1}$ such that

$$\eta_1 = 2\eta_2 \qquad \text{and} \qquad \eta_3 = 0.$$

Prove that the projection of $(\xi_i)_{i\geq 1}$ on $\mathbb{H}_1$ is $P(\xi_i)_{i\geq 1} = (\eta_i)_{i\geq 1}$ given by

$$\eta_i = \begin{cases} \frac{4\xi_1+2\xi_2}{5}, & i = 1, \\ \frac{4\xi_1+2\xi_2}{10}, & i = 2, \\ 0, & i = 3, \\ \xi_i, & i \geq 4. \end{cases}$$

Exercise 9.14. Let ℓ^2_p be the space of Exercise 9.7, and let $\mathbb{H}_1$ be its closed subspace of sequences $(\eta_i)_{i\geq 1} \in \ell^2_p$ such that $\eta_{2i} = \eta_{2i-1}, i \geq 1$. Find a formula for the projection on $\mathbb{H}_1$.

Exercise 9.15. Let $\mathbb{H}_1 \subset L^2[0,1]$ be composed of square integrable functions on the unit interval $[0,1]$ that are constant on $[0, \frac{1}{2}]$. Check that this is a closed subspace and find the formula for the orthogonal projection on this subspace.

☞ CHAPTER SUMMARY

A Hilbert space is a specific example of a Banach space, because its norm comes from a scalar product. This particular norm makes the geometry of a Hilbert space very familiar to us. In particular, in a Hilbert space one can find a unique element of a closed, convex subset that minimizes the distance of this subset from a point lying outside of it. One can also think of projections of vectors on closed subspaces. Again, all this would have been impossible were the space with the scalar product not complete. The chapter ends with a remarkable example showing that conditional probability, one of the fundamental notions of probability theory, has much to do with projections in a Hilbert space.

10
Complete Orthonormal Sequences

The situation when the subspace on which we project is spanned by a system of orthonormal vectors is by far the most important and leads to quite elegant theory. It is one of the consequences of this theory, discussed briefly in this chapter, that the spaces $L^2[a,b]$ and ℓ^2, which seem to be rather different, are in fact quite the same. With this information under our belt, in the next chapter we will with ease solve the heat equation.

10.1 Orthonormal Systems

10.1 Definition We say that a sequence $(e_n)_{n\geq 1}$ of elements of a Hilbert space $\mathbb{H}$ forms an orthonormal system (or, less precisely, that these vectors are orthonormal) if

$$(e_n, e_m) = \begin{cases} 1, & n = m, \\ 0, & n \neq m. \end{cases}$$

For example, it is easy to see that

$$e_n := (0,0,\ldots,0, \underbrace{1}_{n\text{th coordinate}} ,0,\ldots) \in \ell^2, \qquad n \geq 1, \tag{10.1}$$

defines an orthonormal system. Here is a more interesting sequence $(e_n)_{n\geq 1}$, this time in the space $L^2[-\pi,\pi]$:

$$e_n(s) = \frac{1}{\sqrt{\pi}} \sin ns, \qquad s \in [-\pi,\pi], n \geq 1. \tag{10.2}$$

Indeed, since in this space $(x,y) = \int_{-\pi}^{\pi} x(s)y(s)\,\mathrm{d}s$, we have

$$(e_n, e_m) = \frac{1}{\pi}\int_{-\pi}^{\pi} \sin ns \sin ms \,\mathrm{d}x = \frac{1}{2\pi}\int_{-\pi}^{\pi} [\cos(n-m)s - \cos(n+m)s]\,\mathrm{d}s.$$

The last integral is zero provided that $n \neq m$. For $m = n$, the integral is 2π and the entire expression reduces to 1. This shows that $(e_n)_{n\geq 1}$ is an orthonormal sequence in $L^2[-\pi, \pi]$.

To proceed, let $(e_n)_{n\geq 1}$ be a given orthonormal sequence. We fix $m \geq 1$ and consider the subspace $\mathbb{H}_1$ of vectors of the form

$$y = \sum_{n=1}^{m} \alpha_n e_n,$$

that is, of linear combinations of vectors $e_1, \dots, e_m$ – this subspace is often said to be spanned by $e_1, \dots, e_m$. What does the projection Px on $\mathbb{H}_1$ of a vector $x \in \mathbb{H}$ look like? Since $x - Px$ is to be perpendicular to all elements of $\mathbb{H}_1$, it must in particular be perpendicular to all $e_n, n = 1, \dots, m$. Thus, if $Px = \sum_{i=1}^{m} \alpha_i e_i$, then we must have

$$0 = (x - Px, e_n) = (x, e_n) - \alpha_n,$$

where, in the last step, we used the fact that $(e_n)_{n\geq 1}$ forms an orthonormal system. In other words

$$Px = \sum_{n=1}^{m} (x, e_n) e_n;$$

this is an elegant formula that is worth remembering.

We also note that if Px is a projection of x on a subspace $\mathbb{H}_1$, then $x - Px$ is orthogonal to Px. Therefore, by the theorem of Pythagoras,

$$\|x\|^2 = \|x - Px\|^2 + \|Px\|^2,$$

and this implies

$$\|Px\|^2 \leq \|x\|^2.$$

The same theorem applied $(m - 1)$ times (or a direct calculation) proves that

$$\left\| \sum_{n=1}^{m} (x, e_n) e_n \right\|^2 = \sum_{n=1}^{m} [(x, e_n)]^2.$$

Combining these facts we obtain the following important lemma.

10.2 Lemma (Bessel's inequality) *Suppose $(e_n)_{n\geq 1}$ is an orthonormal system. Then, for all $x \in \mathbb{H}$,*

$$\sum_{n=1}^{\infty} [(x, e_n)]^2 \leq \|x\|^2.$$

Proof We have in fact provided the proof already: we have shown that

$$\sum_{n=1}^{m}[(x,e_n)]^2 \le \|x\|^2 \qquad \text{for all } m \ge 1 \text{ and } x \in \mathbb{H}.$$

This is just a different formulation of Bessel's inequality. □

It is natural to ask if Bessel's inequality can be replaced by equality for all $x \in \mathbb{H}$. The answer is: sometimes yes, sometimes no. In the first case we can conclude that there are sufficiently many vectors e_n; that, in other words, the system they form is sufficiently rich, sufficiently large – that is, complete (see also below). We have used the term 'complete' in this book already in a different context, in the context that is fundamental for this book. Notwithstanding the fact that completeness of an orthonormal system has little to do with completeness of a normed space, both notions describe a somewhat similar situation in which the space or the system 'lacks nothing,' that is, is perfect.

10.3 Definition An orthonormal $(e_n)_{n\ge1}$ system in a Hilbert space $\mathbb{H}$ is said to be complete if for all $x \in \mathbb{H}$, Bessel's inequality can be replaced by equality. The resulting formula

$$\|x\|^2 = \sum_{n=1}^{\infty}[(x,e_n)]^2, \qquad x \in \mathbb{H},$$

is referred to as Parseval's identity.[1]

Let's come to the example showing that – quite understandably – even infinite systems of orthogonal vectors need not be complete. Let's think of ℓ^2 and of the sytem $\left(e'_n\right)_{n\ge1}$ where $e'_n = e_{3n}$ and $(e_n)_{n\ge1}$ is defined by (10.1). Clearly, this is an orthogonal system, as a subsystem of a larger orthogonal system. For this system and all $x = (\xi_i)_{i\ge1}$, Bessel's inequality takes the form

$$\sum_{n=1}^{\infty}\xi_{3n}^2 \le \sum_{n=1}^{\infty}\xi_n^2,$$

and it is clear that in general the inequality cannot be replaced by equality. More precisely, equality holds merely for those x having all coordinates zero, except perhaps for those with indexes divisible by 3. A similar reasoning shows, though, that $(e_n)_{n\ge1}$ is complete. Thus, the system $\left(e'_n\right)_{n\ge1}$ is not complete because it was obtained by throwing away some elements of $(e_n)_{n\ge1}$.

[1] Marc-Antoine Parseval des Chênes (1755–1836) should not be confused with Parsifal, son of Pellinore, a Knight of the Round Table.

Here is a less obvious example. Let's consider the space $L^2[-\pi,\pi]$ and the orthogonal system (10.2). Is this system complete? For x defined by $x(s) = 1, s \in [-\pi,\pi]$ and all n, we have

$$(x, e_n) = 0$$

because sine is an odd function. The left-hand side of Bessel's inequality thus equals 0, but the right-hand side is $\int_{-\pi}^{\pi} 1 \,\mathrm{d}x = 2\pi$, and equality cannot hold. This shows that the system is not complete. By the way, this argument works for any even function x. The system (10.2) can be completed by attaching to it the following functions:

$$s \mapsto \frac{1}{\sqrt{2\pi}}, s \mapsto \frac{1}{\sqrt{\pi}} \cos s, s \mapsto \frac{1}{\sqrt{\pi}} \cos 2s, \ldots \tag{10.3}$$

(note that the first constant differs from the remaining constants; this is not a mistake). It is a fundamental statement of the theory of Fourier series that the so-extended system is complete. In particular, for any square integrable function x we have Parseval's identity:

$$\int_{-\pi}^{\pi} [x(s)]^2 \,\mathrm{d}s = \frac{1}{\pi} \sum_{n=1}^{\infty} \left(\int_{-\pi}^{\pi} x(s) \sin ns \,\mathrm{d}s \right)^2 + \frac{1}{2\pi} \left(\int_{-\pi}^{\pi} x(s) \,\mathrm{d}s \right)^2 + \frac{1}{\pi} \sum_{n=1}^{\infty} \left(\int_{-\pi}^{\pi} x(s) \cos ns \,\mathrm{d}x \right)^2. \tag{10.4}$$

On the left, we have the squared norm of function x; on the right, the sum of its squared Fourier coefficients, that is, of the squared scalar products (x, e_n).

To prove that the functions defined in (10.2) and (10.3) indeed form a complete system, we need the following theorem, which provides equivalent conditions for completeness of an orthogonal system.

10.4 Theorem *Let $(e_n)_{n\geq 1}$ be an orthonormal system in a Hilbert space $\mathbb{H}$. The following conditions are equivalent.*

(a) $(e_n)_{n\geq 1}$ is complete.
(b) For any $x \in \mathbb{H}$,

$$x = \sum_{n=1}^{\infty} (x, e_n) e_n \left(:= \lim_{m\to\infty} \sum_{n=1}^{m} (x, e_n) e_n \right).$$

(c) The only x with the property that $(x, e_n) = 0$ for all n is $x = 0$.

(d) *The set of finite linear combinations of $e_n, n \geq 1$ is dense in $\mathbb{H}$; that is, for any $x \in \mathbb{H}$ and any $\epsilon > 0$, there are an integer n and numbers $a_1, \ldots, a_n$ such that $\|x - \sum_{\ell=1}^{n} a_\ell e_\ell\| < \epsilon$.*

Proof Let's show first that (a) implies (c). To this end, suppose that for a certain x, all scalar products (x, e_n) are zero. By the Parseval identity, then, the norm of x is zero, and thus x is a zero vector.

To prove that (c) implies (b) we note that, regardless of whether the system is complete or not, the series

$$\sum_{n=1}^{\infty} (x, e_n) e_n$$

converges. Indeed, the related partial sums $s_m = \sum_{n=1}^{m} (x, e_n) e_n, m \geq 1$ satisfy

$$\|s_m - s_k\|^2 = \|\sum_{n=k+1}^{m} (x, e_n)\|^2 = \sum_{n=k+1}^{m} [(x, e_n)]^2, \qquad k < m.$$

Since, by the Bessel inequality, the series $\sum_{n=1}^{\infty} [(x, e_n)]^2$ converges, these sums form a Cauchy sequence, proving our claim by completeness of $\mathbb{H}$.

Next, let

$$y := x - \sum_{n=1}^{\infty} (x, e_n) e_n.$$

For all $n \geq 1$ we have

$$(y, e_n) = \lim_{m \to \infty} (x - s_m, e_n) = (x, e_n) - \lim_{m \to \infty} (s_m, e_n) = (x, e_n) - (x, e_n) = 0.$$

Assumption (b) then forces $y = 0$, that is, $x = \sum_{n=1}^{\infty} (x, e_n) e_n$.

It is now time to prove that (b) $\Rightarrow$ (a). To this end we recall that for any sequence $(x_n)_{n \geq 1}$, $x = \lim_{n \to \infty} x_n$ implies (by the triangle inequality) $\|x\| = \lim_{n \to \infty} \|x_n\|$, and, of course, $\|x\|^2 = \lim_{n \to \infty} \|x_n\|^2$. Under assumption (b), condition (a) is a direct consequence of this fact.

To summarize: we have proved that (a) $\Leftrightarrow$ (b) $\Leftrightarrow$ (c). Since it is clear that (b) implies (d), we are left with showing that (d) $\Rightarrow$ (c). Suppose therefore that (d) holds and that an $x \in \mathbb{H}$ is such that $(x, e_n) = 0$ for all $n \geq 1$. Then, given $\epsilon > 0$ we can, by (d), find $x_n := \sum_{\ell=1}^{n} a_\ell e_\ell$ such that $\|x - x_n\| < \epsilon$. Then

$$\|x\|^2 = (x, x) = (x, x - x_n) + (x, x_n) = (x, x - x_n) \leq \|x\| \, \|x - x_n\| \leq \epsilon \|x\|.$$

We have thus $\|x\| = 0$ because $\epsilon > 0$ is arbitrary. This establishes (c) and completes the proof. □

To show that functions defined in (10.2) and (10.3) form a complete orthonormal system, we use Corollary 7.8 and the fact, which we take for granted in this book,[2] that the space $C_{\mathrm{p}}[-\pi,\pi]$ is dense in $L^2[-\pi,\pi]$. To recall, Corollary 7.8 says that any $x \in C_{\mathrm{p}}[-\pi,\pi]$ can be uniformly approximated by trigonometric polynomials. That is, for any $x \in C_{\mathrm{p}}[-\pi,\pi]$ and $\epsilon > 0$ there exist an integer n and numbers[3] $a_0,\ldots,a_n$ and $b_1,\ldots,b_n$ such that for p given by

$$p(t) = a_0 + \sum_{\ell=1}^{n} (a_\ell \cos \ell t + b_\ell \sin \ell t),$$

we have

$$\|x - p\|_{C_{\mathrm{p}}[-\pi,\pi]} = \max_{t\in[-\pi,\pi]} |x(t) - p(t)| < \epsilon. \tag{10.5}$$

Combining these two ingredients with the fact that each member of $C_{\mathrm{p}}[-\pi,\pi]$ belongs also to $L^2[-\pi,\pi]$, and

$$\|x\|_{L^2[-\pi,\pi]} = \sqrt{\int_{-\pi}^{\pi} x^2(s)\,\mathrm{d}s} \le \sqrt{2\pi} \max_{s\in[-\pi,\pi]} |x(s)| = \sqrt{2\pi}\,\|x\|_{C_p[-\pi,\pi]},$$

we argue as follows. For a $y \in L^2[-\pi,\pi]$ and $\epsilon > 0$ we first find an $x \in C_{\mathrm{p}}[-\pi,\pi]$ such that

$$\|y - x\|_{L^2[-\pi,\pi]} < \frac{\epsilon}{2},$$

and then a trigonometric polynomial p such that (10.5) is true with ϵ replaced by $\frac{\epsilon}{2\sqrt{2\pi}}$. Then

$$\begin{aligned}\|y - p\|_{L^2[-\pi,\pi]} &\le \|y - x\|_{L^2[-\pi,\pi]} + \|x - p\|_{L^2[-\pi,\pi]}\\ &\le \|y - x\|_{L^2[-\pi,\pi]} + \sqrt{2\pi}\,\|x - p\|_{C_p[-\pi,\pi]} < \epsilon,\end{aligned}$$

proving that (d) in Theorem 10.4 is satisfied. That is, trigonometric polynomials form a complete orthonormal system.

In the case of the orthonormal system (10.2) completed by attaching to it the vectors (10.3), the series of point (b) in Theorem 10.4 is customarily rearranged and written as follows:

[2] For those readers that are familiar with Lebesgue integration: the point is that characteristic functions of open subintervals of $[-\pi,\pi]$ can be approximated by continuous functions (in the norm of $L^2[-\pi,\pi]$). This is just one step away from concluding that the characteristic functions of all measurable subsets of $[-\pi,\pi]$ can be approximated by continuous functions. Linear combinations of characteristic functions in turn are, by the definition of the Lebesgue integral and a little thought, dense in $L^2[-\infty,\infty]$.

[3] These numbers need not coincide with the Fourier coefficients defined later on in (10.7).

$$x(t) = \frac{a_0}{2} + \sum_{n=1}^{\infty} (a_n \cos nt + b_n \sin nt), \tag{10.6}$$

where

$$a_n := \frac{1}{\pi} \int_{-\pi}^{\pi} x(s) \cos ns \,\mathrm{d}s \quad \text{and} \quad b_n := \frac{1}{\pi} \int_{-\pi}^{\pi} x(s) \sin ns \,\mathrm{d}s, \qquad n \geq 1. \tag{10.7}$$

The series on the right-hand side of (10.6) is referred to as the *Fourier series* for x.

It should be stressed that (10.6) *does not* say that for all or any $t \in [-\pi, \pi]$ the numerical series on the right converges to $x(t)$. In the light of Theorem 10.4, all we can say is that the sequence $(s_n)_{n \geq 1}$ of partial sums of the Fourier series defined by

$$s_n(t) := \frac{a_0}{2} + \sum_{\ell=1}^{n} (a_\ell \sin \ell t + b_\ell \cos \ell t), \qquad t \in [-\pi, \pi],$$

converges to x in the norm of $L^2[-\pi, \pi]$:

$$\lim_{n \to \infty} \|s_n - x\|_{L^2} = 0.$$

10.2 Isometric Isomorphism of ℓ^2 and L^2

Let's pause here. We have reached this mountain peak so fast we forgot to notice the view we have right before our eyes. And indeed we have proved something that is definitely worth our attention. We have proved something unexpected. Can you see that yet?

We have proved that if in a Hilbert space there is an orthonormal sequence that is complete, then this space is indistinguishable from ℓ^2.

Let's look through field glasses. Suppose $(e_n)_{n \geq 1}$ is a complete orthonormal system in $\mathbb{H}$. Then to each $x \in \mathbb{H}$ we can assign the sequence $\xi_i = (x, e_i), i \geq 1$, and Parseval's identity shows that $(\xi_i)_{i \geq 1}$ belongs to ℓ^2. The map

$$\mathbb{H} \ni x \mapsto Ix = (\xi_i)_{i \geq 1} \in \ell^2$$

is linear, that is, $I(\alpha x + \beta y) = \alpha Ix + \beta Iy$, so that the linear structure in ℓ^2 is a counterpart of the linear structure in $\mathbb{H}$. This map is also onto, because given a sequence $(\xi_i)_{i \geq 1}$ in ℓ^2, we can define an x in $\mathbb{H}$ by (see the proof of Theorem 10.4)

$$x = \sum_{n=1}^{\infty} \xi_n e_n,$$

and then, as it is easy to see, $Ix = (\xi_i)_{i\geq 1}$. Additionally, since $(e_n)_{n\geq 1}$ is complete, Parseval's identity yields

$$\|x\|_{\mathbb{H}}^2 = \|Ix\|_{\ell^2}^2,$$

and this says that the norms in $\mathbb{H}$ and ℓ^2 are 'the same.' In particular, I is one-to-one: $Ix = 0$ implies $\|Ix\|_{\ell^2} = 0$, that is, $\|x\|_{\mathbb{H}} = 0$ and so, consequently, $x = 0$.

Maps that, like I, preserve the linear structure and the norm are said to be isometric isomorphisms of Hilbert or Banach spaces. And two Hilbert or Banach spaces are said to be isometrically isomorphic provided that there is an isometric isomorphism between them. Hence, our result can be formally phrased as follows: if in a Hilbert space there is a complete orthonormal sequence, then this space is isometrically isomorphic to ℓ^2.

You are not impressed with this statement yet? Then, let's state its immediate consequence, forced by the fact that the sequence of sines and cosines is complete in $L^2[-\pi,\pi]$.

> The space $L^2[-\pi,\pi]$ is isometrically isomorphic to ℓ^2.

What? Really? How come?

10.3 Exercises

Exercise 10.1. Expand the function $x(t) = t, t \in [-\pi,\pi]$ into a Fourier series. Take $t = \pi$ in this expansion to see that the obtained series does not sum to $x(\pi)$ but to 0. (We thus convince ourselves once again that convergence in $L^2[-\pi,\pi]$ is not pointwise convergence.) Write down Parseval's identity for this x.

Exercise 10.2. Convergence of a numerical series implies convergence to zero of its terms. Use (10.4) to prove the following version of Riemann's lemma (Exercise 7.7): for any $x \in L^2[-\pi,\pi]$,

$$\lim_{n\to\infty} \int_{-\pi}^{\pi} x(s) \sin ns \, ds = \lim_{n\to\infty} \int_{-\pi}^{\pi} x(s) \cos ns \, ds = 0.$$

Exercise 10.3. Orthogonal polynomials, such as the polynomials of Legendre, Hermite and Laguerre, lead to important examples of orthonormal systems (see [27]). In this exercise we briefly cover the case of Hermite polynomials. They are defined by

$$H_n(t) = (-1)^n e^{t^2} \left(e^{-t^2}\right)^{(n)}, \qquad n \geq 0,$$

where (n) denotes the nth order derivative with respect to variable t.

1. Check by direct calculation that

$$H_0(t) = 1,\ H_1(t) = 2t,\ H_2(t) = 4t^2 - 2 \quad \text{and} \quad H_3(t) = 8t^3 - 12t.$$

2. Prove that

$$H_{n+1}(t) = 2t H_n(t) - 2n H_{n-1}(t), \qquad n \geq 1, \tag{10.8}$$

and use this recurrence to find a couple of consecutive polynomials ($n \geq 4$).

3. Convince yourself that

$$H_n'(t) = 2n H_{n-1}(t),$$

and (by (10.8))

$$H_{n+1}(t) = 2t H_n(t) - H_n'(t), \qquad n \geq 1. \tag{10.9}$$

4. Conclude that H_n is a solution to the equation

$$u'' - 2tu' + 2nu = 0. \tag{10.10}$$

5. Let $L^2(e^{-t^2/2})$ denote the space of functions that after being multiplied by the weight $t \mapsto w(t) = e^{-t^2/2}$ become square integrable on the entire real line. That means that a ϕ belongs to $L^2(e^{-t^2/2})$ if $t \mapsto w(t)\phi(t)$ is square integrable, and

$$\|\phi\|^2_{L^2(e^{-t^2/2})} = \int_{\mathbb{R}} \phi^2(t) e^{-t^2}\, dt.$$

Convince yourself (following our analysis of Chapter 5, where we considered the algebra $L^1_\omega(\mathbb{R}^+)$) that this is a Hilbert space that is not distinguishable from $L^2(\mathbb{R})$. What is the formula for the scalar product in this space?

6. Check to see that Hermite polynomials are elements of $L^2(e^{-t^2/2})$ and form an orthogonal sequence in $L^2(e^{-t^2/2})$, that is,

$$\int_{\mathbb{R}} H_n(t) H_m(t) e^{-t^2}\, dt = 0, \qquad n \neq m.$$

Hint: Use point 4 to prove that, if $G_n := wH_n, n \geq 0$, then for all m and n,

$$G_n'' + (2n + 1 - t^2)G_n = 0 \qquad \text{and} \qquad G_m'' + (2m + 1 - t^2)G_m = 0.$$

Then, by multiplying the first equation by G_m and multiplying the second equation by G_n, check to see that

$$(G_n' G_m - G_m' G_n)' + 2(n - m)G_n G_m = 0,$$

and finally integrate both sides over $\mathbb{R}$.

Exercise 10.4. Find a natural isometric isomorphism between the space ℓ^2 of square summable sequences and the space ℓ_p^2 of Exercise 9.7. What is the counterpart in ℓ^2 of the subspace $\mathbb{H}_1$ defined in this exercise? Find a projection on the subspace of ℓ^2 that is a counterpart of $\mathbb{H}_1$. How does the result relate to the projection found in Exercise 9.7?

☞ CHAPTER SUMMARY

A sequence of norm-one elements of a Hilbert space that are mutually orthogonal is said to form an orthonormal sequence. If, additionally, such a sequence spans the entire space, it is said to be complete. As it turns out, if in a Hilbert space there is a complete orthonormal sequence, this space is indistinguishable from ℓ^2. In particular, perhaps contrary to our misleading intuition saying that there are many more square integrable functions than there are square summable sequences, $L^2[-\pi,\pi]$ is as large as (in fact much the same as) ℓ^2. We will see one important consequence of this stunning result in the next chapter.

11
Heat Equation

In this chapter, using the theory developed in Chapter 10, we will solve the heat equation, one of the most important equations of mathematical physics.

11.1 A Convenient Complete System in $L^2[0,\pi]$

In the surprising theorem of the previous chapter, it is not important that the interval's length is 2π. All spaces $L^2[a,b]$ $(a < b)$ are isometrically isomorphic, and thus isometrically isomorphic to ℓ^2 (see Exercise 11.3). In particular, this applies to $L^2[0,\pi]$, even though this space looks like a half of $L^2[-\pi,\pi]$. We will use this fact to solve the heat equation, but first we need to find a convenient complete system there (there are, of course, many complete systems in $L^2[0,\pi]$; the point is to find a *convenient* one).

We will show, namely that $(e_n)_{n\geq 1}$ defined by

$$e_n(s) = \sqrt{\frac{2}{\pi}}\sin ns, \qquad s \in [0,\pi], n \geq 1, \tag{11.1}$$

is convenient indeed. (We leave it to the reader to check that this is an orthonormal system.) We remark here that we have previously proved that a similar system (but differently scaled) is not complete in $L^2[-\pi,\pi]$. Now we will show that in $L^2[0,\pi]$ it is complete. (Didn't I say that $L^2[0,\pi]$ is just a half of $L^2[-\pi,\pi]$? In the latter space, besides sines you must also have cosines.)

To this end, we will use Theorem 10.4. Suppose namely that $x \in L^2[0,\pi]$ and all scalar products (x,e_n) are zero. Let $\tilde{x}$ be an odd extension of x to $[-\pi,\pi]$, that is, let $\tilde{x}(-s) = -x(s), s \in (0,\pi]$. Since cos is even,

$$\int_{-\pi}^{\pi} \tilde{x}(s)\cos ns \, \mathrm{d}s = 0, \qquad n \geq 0.$$

On the other hand, sine being odd, all functions $s \mapsto \tilde{x}(s)\sin ns, n \geq 1$ are even, and we have

$$\int_{-\pi}^{\pi} \tilde{x}(s) \sin ns \, ds = 2\int_{0}^{\pi} x(s) \sin ns \, ds = 0$$

(the last equality by assumption). Therefore, completeness of the system composed of sines and cosines in $L^2[-\pi,\pi]$ implies $\tilde{x} = 0$ and $x = 0$ (almost everywhere). This in turn establishes the following proposition.

11.1 Theorem *The system* (11.1) *is complete.*

As promised, in the following sections, we will use this information to solve the heat equation (and explain in what sense this system in convenient).

11.2 Problem Statement

Imagine a short, but very, very thin wire – we do not think of a piece of wood or plastic because we want the material to conduct heat well. Imagine also that at time $t = 0$ we know temperature distribution across the wire, and would like to predict this distribution's shape in the future. Additionally, we assume that the temperature at the wire's ends is constant, and for convenience we set it to equal zero. It turns out that, from the mathematical point of view, this leads to the following problem: having given a function $x_0\colon [0,\pi] \to \mathbb{R}$ that models the initial temperature's distribution,[1] find a function $u\colon \mathbb{R}^+ \times [0,\pi]$ of two variables, that models the evolution of temperature's distribution in time, such that

(a) the following relation, known as the heat equation, holds:[2]

$$\frac{\partial u(t,s)}{\partial t} = \frac{\partial^2 u(t,s)}{\partial s^2}, \qquad s \in [0,\pi], t \geq 0, \tag{11.2}$$

(at the ends of the interval, the derivates are one-sided);

(b) the following conditions, known as Dirichlet boundary conditions, are satisfied:

$$u(t,0) = u(t,\pi) = 0, \qquad t \geq 0;$$

[1] As you can see we set the wire's length to be π – we hasten to add that this is for convenience only; a different choice would lead to more complicated formulae, but the core of the reasoning would be the same.

[2] We are cheating, but only slightly. On the right-hand side, an important coefficient, known as the thermal diffusivity, is missing. But we are mathematicians; for us, a coefficient is a constant we can get rid of by slightly changing variables.

(c) the initial condition

$$u(0,s) = x_0(s), \qquad s \in [0,\pi],$$

is satisfied.

This problem, involving one of the most important equations of mathematical physics [19, 32, 41, 42], can be solved in many (basically equivalent) ways. In a moment, we will reformulate it so that we will be able to use the fact that $L^2[0,\pi]$ is isometrically isometric to ℓ^2, but before we do this, we need to take a short course in generalized derivatives.

11.3 Generalized Derivatives; Sobolev Spaces

As already recalled, the fundamental theorem of calculus says that if a function x, defined on an open interval (a,b), is continuously differentiable there, then x can be 'recovered' from x' as follows:

$$x(s) = x(s_0) + \int_{s_0}^{s} x'(\sigma)\,\mathrm{d}\sigma \tag{11.3}$$

for all s_0 and s in this interval. The function $x(s) = |s|$, defined, say, in $(-1,1)$, is not differentiable at $s = 0$, and so for this x the theorem does not apply. Nevertheless,

$$x(s) = x(s_0) + \int_{s_0}^{s} y(\sigma)\,\mathrm{d}\sigma \tag{11.4}$$

for y defined as follows:

$$y(\sigma) = \operatorname{sgn}\sigma = \begin{cases} 1, & \sigma > 0, \\ 0, & \sigma = 0, \\ -1, & \sigma < 0. \end{cases}$$

Obviously, we cannot claim that y is a 'true' derivative of x. We see, however, that y plays a role that is analogous to that of x', while the latter does not even exist.

11.2 Definition If there are an s_0 and an integrable function y such that

$$x(s) = x(s_0) + \int_{s_0}^{s} y(\sigma)\,\mathrm{d}\sigma \tag{11.5}$$

for all s in an interval (a,b), we say that x is *absolutely continuous* in this interval. If this formula holds for $s = a$ and $s = b$ also, we say that x is absolutely continuous in $[a,b]$.

For y satisfying (11.5), we will write

$$y = x', \tag{11.6}$$

remembering, however, that this is a derivative in a generalized sense. We note that by changing the values of y at a point or two (or, as a matter of fact, in a set of measure zero) we do not alter its defining property (11.5). In other words, (11.6) determines y merely almost everywhere.

Interestingly, functions that are differentiable in this generalized sense form Hilbert (and Banach) spaces, known under the common name of Sobolev spaces – the latter are of fundamental importance in the theory of partial differential equations (see e.g., [2]). For example, the space $H_0^1(a,b]$, which is composed of functions that vanish at $t = a$ and are absolutely continuous in this interval with $x' \in L^2[a,b]$, is a Hilbert space. It is quite easy to see (see Exercise 11.5) that the formula

$$(x_1, x_2) = \int_a^b x_1'(s) x_2'(s) \, \mathrm{d}s \tag{11.7}$$

describes the scalar product in $H_0^1(a,b]$; the integral on the right-hand side makes sense because both x_1' and x_2' belong to $L^2[a,b]$. This formula can be written as follows:

$$(x_1, x_2)_{H_0^1(a,b]} = (x_1', x_2')_{L^2[a,b]},$$

where the indexes serve to distinguish the scalar product in $H_0^1(a,b]$ from that in $L^2[a,b]$. In particular,

$$\|x\|_{H_0^1(a,b]} = \|x'\|_{L^2[a,b]}. \tag{11.8}$$

In fact, it is quite easy to show that $H_0^1(a,b]$ is complete. To see this, assume that $(x_n)_{n\geq 1}$ is a Cauchy sequence in this space. By the definition of the norm in $H_0^1(a,b]$, we see that $\left(x_n'\right)_{n\geq 1}$ is a fundamental sequence in $L^2[a,b]$; therefore, it has a limit, say y. Then the formula $x(s) = \int_a^s y(\sigma)\, \mathrm{d}\sigma, s \in [a,b]$, defines an element of $H_0^1(a,b]$, and this x is a natural candidate for the limit of $(x_n)_{n\geq 1}$. To prove that x is indeed the limit we seek, we check that

$$\|x_n - x\|_{H_0^1(a,b]} = \|x_n' - x'\|_{L^2[a,b]} = \|x_n' - y\|_{L^2[a,b]},$$

and note that the last expression converges to 0, as $n \to \infty$.

We end this section by noting that, since for each $x \in H_0^1(a,b]$ we have $x(s) = \int_a^s x'(\sigma)\,\mathrm{d}\sigma, s \in [a,b]$, we can estimate as follows:

$$|x(s)| \leq \int_a^s |x'(\sigma)|\,\mathrm{d}\sigma \leq \int_a^b |x'(\sigma)| \cdot 1\,\mathrm{d}\sigma \leq \sqrt{\int_a^b |x'(\sigma)|^2\,\mathrm{d}\sigma}\sqrt{\int_a^b 1^2\,\mathrm{d}\sigma}$$
$$= \sqrt{b-a}\|x\|_{H_0^1(a,b]},$$

where, in the third step, we used the Cauchy–Schwarz inequality in $L^2[a,b]$. It follows that

$$\|x\|_{C[a,b]} = \sup_{s\in[a,b]} |x(s)| \leq \sqrt{b-a}\|x\|_{H_0^1(a,b]};$$

this is a simple case of the celebrated Sobolev inequalities [2].

As a direct consequence, we see that a sequence that converges in the norm of $H_0^1(a,b]$ also converges uniformly in $s \in [a,b]$. This is not the case with sequences that converge in the norm of $L^2[a,b]$ – convergence in the latter space does not even imply pointwise convergence (see Section 11.5 and Exercise 11.6).

11.4 Definition of Solution

Let's clarify what we will mean by a solution to the heat equation. First of all, instead of thinking of it as a partial differential equation, we will prefer to think of it as an ordinary differential equation, but that in a Banach space $L^2[0,\pi]$. Thus we view (11.2) as

$$u'(t) = Au(t), \qquad u(0) = x_0, \tag{11.9}$$

where x_0 is a given element of $L^2[0,\pi]$, and A is the second derivative operator. More precisely, A is an operator (that is, a map that transforms functions into functions) with domain $\mathcal{D}(A)$ composed of functions x that possess the following properties. First of all, each $x \in \mathcal{D}(A)$ is differentiable in the usual sense (as a function of s); second, x' is absolutely continuous and its generalized derivative x'' belongs to $L^2[0,\pi]$; finally, $x(0) = x(\pi) = 0$. We note that the boundary condition for the heat equation has thus been hidden in the domain of A. Of course, we also agree that $Ax = x''$.

Such a statement of the problem changes our perspective considerably. We are not looking for a function of two variables $t \geq 0$ and $s \in [0,\pi]$, but rather for an $L^2[0,\pi]$-valued function of one variable $t \geq 0$. To be more precise, we want to find a map $(0, \infty) \ni t \mapsto u(t) \in L^2[0, \pi]$ with the following properties:

(a) For each $t > 0$, $u(t)$ belongs to $\mathcal{D}(A)$ (otherwise the right-hand side of (11.9) makes no sense).
(b) Function $(0, \infty) \ni t \mapsto u(t)$ is continuously differentiable (in the sense of the norm of $L^2[0, \pi]$). This means that, for each $t > 0$, there is a $u'(t) \in L^2[0, \pi]$ such that

$$\lim_{h \to 0} \left\| \frac{u(t+h) - u(t)}{h} - u'(t) \right\| = 0, \tag{11.10}$$

and the function $t \mapsto u'(t)$ is continuous (again, this is continuity in the sense of $L^2[0, \pi]$).
(c) The limit $\lim_{t \to 0+} u(t)$ exists and equals x_0 – this is a counterpart of the initial condition.
(d) For each $t > 0$, we have $u'(t) = Au(t)$.

We again stress the fact that equations (11.9) and (11.2), though apparently similar, in fact differ considerably. In particular, we recall that convergence in $L^2[0, \pi]$ does not imply pointwise convergence, and vice versa (see Exercises 10.1 and 11.6). Thus, the derivative of point (b) should not be, a priori, identified with the partial derivative $\frac{\partial u}{\partial t}$. It is a mathematical miracle that – as we shall see later – by solving (11.9) we obtain solutions of (11.2) as well.

If the problem posed above appears to be difficult, it is mostly because we are working in a somewhat uncomfortable space $L^2[0, \pi]$. In the next section, we pass to a more amiable space ℓ^2 where the solution will turn out to be quite natural and simple.

11.5 Solution: How to Do It More Easily

To recall – as we have proved at the beginning of this chapter, $L^2[0, \pi]$ is indistinguishable from ℓ^2, that is, isometrically isomorphic to ℓ^2. By assigning to an x from $L^2[0, \pi]$ the sequence of its Fourier coefficients $(a_n)_{n \ge 1}$

$$a_n := (x, e_n) = \sqrt{\frac{2}{\pi}} \int_0^\pi x(s) \sin ns \, \mathrm{d}s,$$

we obtain an element of ℓ^2. And vice versa, to a sequence $(a_n)_{n \ge 1}$ from ℓ^2, there corresponds the

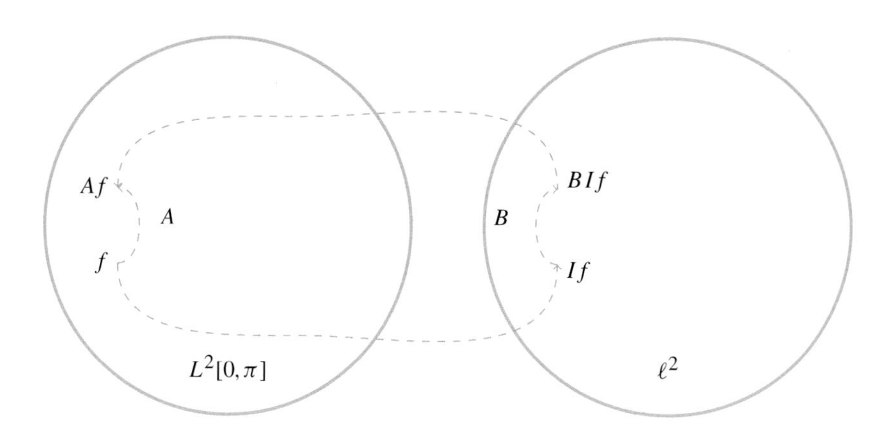

Figure 11.1 In ℓ^2, B plays the role of A.

$$x := \sum_{n=1}^{\infty} a_n e_n,$$

where $e_n(s) = \sqrt{\frac{2}{\pi}} \sin ns, s \in [0,\pi]$. Furthermore, the norm of x in $L^2[0,\pi]$ is the same as the norm of $(a_n)_{n\geq 1}$ in ℓ^2. Since ℓ^2 is more amiable to us, it would be desirable to find a counterpart of (11.9) in this nicer space.

To this end, we introduce the operator B in ℓ^2 given by

$$B(a_n)_{n\geq 1} = -\left(n^2 a_n\right)_{n\geq 1},$$

with domain $\mathcal{D}(B)$ composed of all $(a_n)_{n\geq 1}$ such that $\left(n^2 a_n\right)_{n\geq 1}$ belongs to ℓ^2. For example, any sequence with all but a finite number of coordinates equal to zero belongs to $\mathcal{D}(B)$, and so does $\left(\frac{1}{n^3}\right)_{n\geq 1}$; however $\left(\frac{1}{n^2}\right)_{n\geq 1}$ does not. Even though operator B looks much simpler than A, they are – as we shall see now – in fact quite identical.

11.3 Lemma *B is a counterpart of A in ℓ^2. This means that if we denote by I the operator that assigns the series of Fourier coefficients to an $x \in L^2[0,\pi]$, then*

$$x \in \mathcal{D}(A) \quad \textit{if and only if} \quad Ix \in \mathcal{D}(B),$$

and we have (see Figure 11.1)

$$Ax = I^{-1}BIx \quad \textit{for} \quad x \in \mathcal{D}(A).$$

Proof Each $x \in \mathcal{D}(A)$ is of the form

$$x(s) = \int_0^s \left(c + \int_0^\sigma y(\varsigma)\,\mathrm{d}\varsigma \right) \mathrm{d}\sigma; \qquad s \in [0,\pi],$$

where $y \in L^2[0,\pi]$ is the second derivative of x, and $c = x'(0)$ is a constant; we have used here the fact that $x(0) = 0$. Let a_n and $b_n, n \ge 1$ be the Fourier coefficients of x and y, respectively. Integrating by parts twice and using $x(0) = x(\pi) = 0$, we see that

$$\begin{aligned} a_n\sqrt{\frac{\pi}{2}} &= \int_0^\pi x(s)\sin ns\,\mathrm{d}s = \int_0^\pi \int_0^s (c + \int_0^\sigma y(\varsigma)\,\mathrm{d}\varsigma)\,\mathrm{d}\sigma \sin ns\,\mathrm{d}s \\ &= \int_0^\pi (c + \int_0^s y(\varsigma)\,\mathrm{d}\varsigma)\frac{\cos ns}{n}\,\mathrm{d}s = -\int_0^\pi y(s)\frac{\sin ns}{n^2}\,\mathrm{d}s \\ &= -\frac{1}{n^2}b_n\sqrt{\frac{\pi}{2}}, \qquad n \ge 1. \end{aligned} \tag{11.11}$$

This means that $a_n = -\frac{1}{n^2}b_n$ for all $n \ge 1$, and so $\left(n^2 a_n\right)_{n\ge1} = -\left(b_n\right)_{n\ge1}$ belongs to ℓ^2, because $(b_n)_{n\ge1}$ is the sequence of Fourier coefficients of $y \in L^2[0,\pi]$. Thus, $Ix = (a_n)_{n\ge1}$ belongs to $\mathcal{D}(B)$, $BIx = -\left(n^2 a_n\right)_{n\ge1} = (b_n)_{n\ge1}$ and $I^{-1}BIx = I^{-1}(b_n)_{n\ge1} = y$, proving the second part of the lemma.

The reasoning presented above shows also that condition $x \in \mathcal{D}(A)$ implies $Ix \in \mathcal{D}(B)$; we are left with showing the converse. To this end, we assume that $Ix = (a_n)_{n\ge1}$ is such that $\left(n^2 a_n\right)_{n\ge1}$ belongs to ℓ^2. Then, we can consider $y := -I^{-1}\left(n^2 a_n\right)_{n\ge1}$ and the $\widetilde{x} \in L^2[0,\pi]$ given by the formula

$$\tilde{x}(s) := \int_0^s (c + \int_0^\sigma y(\varsigma)\,\mathrm{d}\varsigma)\,\mathrm{d}\sigma, \qquad s \in [0,\pi],$$

where $c := -\frac{1}{\pi}\int_0^\pi \int_0^\sigma y(\varsigma)\,\mathrm{d}\varsigma\,\mathrm{d}\sigma$ so that $\widetilde{x}(\pi) = 0 = \widetilde{x}(0)$. It is then easy to see that $\widetilde{x}$ belongs to $\mathcal{D}(A)$. The calculations (11.11) show furthermore that the sequence of Fourier coefficients of $\widetilde{x}$ coincides with that of x: the nth coefficient of $\widetilde{x}$ is the nth coefficient of y divided by $-n^2$, and is thus the same as the nth coefficient of x. This means, however, that x and $\widetilde{x}$ are sums of the same series $\sum_{n=1}^\infty a_n e_n$, proving that $x = \widetilde{x}$. □

Hence, if the role of I may be compared to that of a mirror, then everything that happens in $L^2[0,\pi]$ has its reflection in ℓ^2. And the nice property of this mirror is that the reflection of a body visible in it (i.e., the operator B) is clearer than the body itself (i.e., A).[3]

[3] This is completely different than the observation of Paul of Tarsus, who wrote to the Corinthian believers 'For now we see in a mirror obscurely.'

Coming back to the heart of the matter, the lemma shows that instead of a rather complicated differential equation in $L^2[0,\pi]$ we can solve its simpler counterpart in ℓ^2. If $(0,\infty) \ni t \mapsto u(t) \in L^2[0,\pi]$ is a solution of (11.9), then $(0,\infty) \ni v(t) := Iu(t) \in \ell^2$ solves the equation

$$v'(t) = Bv(t), \tag{11.12}$$

with initial condition $v(0) = Ix_0$, and vice versa. Therefore, we will find v first (because it is easier) and then recover u from v.

To summarize, we are looking for a v with the following properties, which reflect the properties of u.

(a) For each $t > 0$, $v(t)$ belongs to $\mathcal{D}(B)$.

(b) Function $(0,\infty) \ni t \mapsto v(t)$ is continuously differentiable (in the sense of the norm of ℓ^2): for each $t > 0$, there is a $v'(t) \in \ell^2$ such that

$$\lim_{h\to 0} \left\| \frac{v(t+h) - v(t)}{h} - v'(t) \right\| = 0,$$

and the function $t \mapsto v'(t)$ is continuous in the norm of ℓ^2.

(c) The limit $\lim_{t\to 0+} v(t)$ exists and equals Ix_0.

(d) For each $t > 0$, we have $v'(t) = Bv(t)$.

Let's also explain why we think that differential equations in ℓ^2 are easier to solve than those in $L^2[0,\pi]$. The main reason is that we understand convergence in ℓ^2 better. Indeed, for each $i \geq 1$,

$$|a_i| \leq \|(a_n)_{n\geq 1}\|_{\ell^2} = \sqrt{\sum_{n=1}^{\infty} a_n^2},$$

and it follows that convergence in the norm of ℓ^2 implies convergence of coordinates. That is, if $(x_n)_{n\geq 1}$, where $x_n = \left(\xi_{i,n}\right)_{i\geq 1}$ is a sequence of elements of ℓ^2 that converges to an $x = (\xi_i)_{i\geq 1} \in \ell^2$ in the norm (meaning $\lim_{n\to\infty} \|x_n - x\|_{\ell^2} = 0$), then for each $i \geq 1$ we also have $\lim_{n\to\infty} \xi_{i,n} = \xi_i$. Even though the converse is not true (consider e.g., $x_n = e_n$ for e_n defined in (10.1)), we still have a good grasp of what convergence in ℓ^2 means: it is 'a bit more' than convergence of coordinates. In $L^2[0,\pi]$ this intuition completely fails, as convergence in the norm and pointwise convergence are quite 'perpendicular' (see Exercise 11.6).

To see that this has an immediate bearing on the solutions to our differential equation, think of a solution $(0,\infty) \ni t \to v(t) \in \ell^2$ to this equation, and assume that it is of the form

$$v(t) = (a_n(t))_{n\geq 1}\,.$$

The discussion presented above shows that, since this solution is differentiable, so is, for all n, the real-valued function $t \mapsto a_n(t)$. Moreover, for each $t > 0$, $a_n'(t)$ is equal to the nth coordinate of $v'(t)$, that is, to the nth coordinate of $Bv(t) = B\,(a_n(t))_{n\geq 1} = \left(-n^2 a_n(t)\right)_{n\geq 1}$. It follows that the function $t \mapsto a_n(t)$ is a solution of the differential equation

$$a_n'(t) = -n^2 a_n(t),$$

with initial condition $a_n(0) = a_n$, where a_n is the nth coordinate of Ix_0. This in turn tells us that

$$a_n(t) = \mathrm{e}^{-n^2 t} a_n.$$

We have thus quickly and with ease determined the form of solution of our problem in ℓ^2. For reasons explained above, our argument would fail in $L^2[0,\pi]$.

11.6 Solution: Verification

Have we already solved the heat equation (or at least its mirror image in ℓ^2)? Not yet. We have merely proved, using the fact that convergence in the norm of ℓ^2 implies convergence of coordinates, that if a solution of (11.12) exists then it must be of the form

$$v(t) = \left(\mathrm{e}^{-n^2 t} a_n\right)_{n\geq 1}, \qquad t > 0. \tag{11.13}$$

But we cannot a priori assume that a solution exists. An attempt to reverse the argument presented above, in Section 11.5, will also obviously fail at the point where we try to prove that pointwise convergence implies norm convergence. There is no other way: we need to check directly that v defined by (11.13) is a solution to (11.12).

Is condition (a) satisfied? In other words, does $\left(\mathrm{e}^{-n^2 t} a_n\right)_{n\geq 1}$ belong to $\mathcal{D}(B)$ for $t > 0$? The answer is in the affirmative, because convergence of the series $\sum_{n=1}^{\infty} a_n^2$ implies

$$\limsup_{n\to\infty} \sqrt[n]{a_n^2} \leq 1,$$

and this, when combined with

$$\lim_{n\to\infty} \sqrt[n]{n^4 \mathrm{e}^{-2n^2 t}} = \lim_{n\to\infty} \sqrt[n]{n^4} \lim_{n\to\infty} \mathrm{e}^{-2nt} = 0,$$

proves convergence of the series $\sum_{n=1}^{\infty} n^4 \mathrm{e}^{-2n^2 t} a_n^2$ by virtue of Cauchy's criterion of convergence. (Under assumption $x_0 \in \mathcal{D}(B)$ – which, to recall, we do not make – the proof would be even easier: convergence of $\sum_{n=1}^{\infty} n^4 a_n^2$ would imply convergence of $\sum_{n=1}^{\infty} n^4 \mathrm{e}^{-2n^2 t} a_n^2$, since obviously $n^4 \mathrm{e}^{-2n^2 t} a_n^2 \le n^4 a_n^2$.) Conditions (b) and (c) will be established once we prove that

$$\lim_{h\to 0} \left\| \frac{v(t+h)-v(t)}{h} - Bv(t) \right\| = 0 \qquad \text{for each } t > 0.$$

To this end, we introduce

$$\frac{v(t+h)-v(t)}{h} - Bv(t) =: \left(c_{n,h}(t)\right)_{n\ge 1}.$$

Then,

$$c_{n,h}(t) = \mathrm{e}^{-n^2 t} a_n \left(\frac{\mathrm{e}^{-hn^2} - 1}{h} + n^2 \right) = \mathrm{e}^{-n^2 t} n^2 a_n \left(1 - h^{-1} \int_0^h \mathrm{e}^{-n^2 s}\, \mathrm{d}s \right). \tag{11.14}$$

It follows that for each n and $t > 0$, $\lim_{t\to 0} c_{n,h}(t) = 0$, but we aim to prove a bit more than this: we wish to show that $\left(c_{n,h}(t)\right)_{n\ge 1}$ converges to 0 in the norm of ℓ^2.

To this end, we note that, for $h > 0$, $|c_{n,h}(t)| \le \mathrm{e}^{-n^2 t} n^2 a_n$, because in this case the integrand in (11.14) does not exceed 1. For $h < 0$, on the other hand, the integrand is greatest at $s = h$ and equals $\mathrm{e}^{-n^2 h}$; this shows that $|c_{n,h}(t)| \le 2\mathrm{e}^{-n^2 t} n^2 a_n \mathrm{e}^{-n^2 h}$. Therefore, as long as $|h| < \frac{t}{2}$,

$$|c_{n,h}(t)| \le c_n(t) := 2\mathrm{e}^{-n^2 \frac{t}{2}} n^2 a_n,$$

and, as we have seen previously, $\sum_{n=1}^{\infty} [c_n(t)]^2$ is finite.

We are now ready to show that the norm of $\left(c_{n,h}(t)\right)_{n\ge 1}$ tends to 0, as $h \to 0$. Given $\epsilon > 0$ we first choose k so that $\sum_{n=k}^{\infty} [c_n(t)]^2 < \frac{\epsilon}{2}$. Next, we choose a positive $h_0 < \frac{t}{2}$ so that $\sum_{n=1}^{k-1} \left(c_{n,h}(t)\right)^2 < \frac{\epsilon}{2}$ for $|h| < h_0$. For such h, the squared norm of $\left(c_{n,h}(t)\right)_{n\ge 1}$ does not exceed

$$\sum_{n=1}^{k-1} \left(c_{n,h}(t)\right)^2 + \sum_{n=k}^{\infty} \left(c_{n,h}(t)\right)^2 < \frac{\epsilon}{2} + \sum_{n=k}^{\infty} [c_n(t)]^2 < \frac{\epsilon}{2} + \frac{\epsilon}{2} = \epsilon,$$

proving our claim.

Proof of (d) is similar, and we leave it to the reader.

11.4 Remark Please note that above we do not assume that Ix belongs to $\mathcal{D}(B)$. And yet, for all $t > 0$, regardless of how small t is, $v(t)$ does belong to $\mathcal{D}(B)$. This situation is quite typical for partial differential equations of parabolic type (the heat equation is one such example).

Having established that v of (11.13) is a true (and the only) solution to equation (11.12), we come back to equation (11.9). Our analysis shows that its solution is

$$u(t) = I^{-1}v(t) = I^{-1}\left(\mathrm{e}^{-n^2 t}a_n\right)_{n\geq 1} = \sum_{n=1}^{\infty} \mathrm{e}^{-n^2 t} a_n e_n. \tag{11.15}$$

It is worth stressing once again that the series on the right converges in the norm of $L^2[0,\pi]$. It may thus be found surprising that the corresponding function series

$$u(t,s) = \sqrt{\frac{2}{\pi}} \sum_{n=1}^{\infty} \mathrm{e}^{-n^2 t} a_n \sin ns \tag{11.16}$$

converges as well, and the convergence is absolute and even uniform with respect to $s \in [0,\pi]$ (for fixed $t > 0$). To prove this one may use the Weierstrass M-test, because $|\mathrm{e}^{-n^2 t} a_n \sin ns| \leq \mathrm{e}^{-n^2 t} a_n$ and the numerical series $\sum_{n=1}^{\infty} \mathrm{e}^{-n^2 t} a_n$ converges.

Similar reasoning shows that u of (11.16) solves the heat equation understood as a partial differential equation. We omit its details, since presenting them would lead us away from the main subject of the book. We note merely that this apparent miracle is caused by the extraordinary smoothing properties of parabolic partial differential equations.

11.7 Exercises

Exercise 11.1. Prove that the sequence $(e_n)_{n\geq 1}$, where e_n is defined in (10.1), does not converge in ℓ^2.

Exercise 11.2. Calculate (x, e_n) for x given by $x(s) = \mathrm{e}^s, s \in [0,\pi]$ and e_n defined in (11.1). The series $\sum_{n=1}^{\infty}(x,e_n)e_n$ is known as the expansion into a series of sines. Repeat the calculation for $x(s) = s$.

Exercise 11.3. Prove that all spaces $L^2[a,b]$, where $-\infty < a < b < \infty$, are isometrically isomorphic to $L^2[0,1]$, and thus isometrically isomorphic to each other. **Hint:** Consider $I\colon L^2[a,b] \to L^2[0,1]$ defined by

$$Ix(s) = \sqrt{b-a}\, x(\chi(s)), \qquad s \in [0,1],$$

where $\chi(s) = (b-a)s + a, s \in [0,1]$.

Exercise 11.4. Check to see that the function depicted in Figure 11.2 is absolutely continuous, even though it is not differentiable at the points $x = 0, 1, 3, 4$ and 6. Write the formula for its generalized derivative.

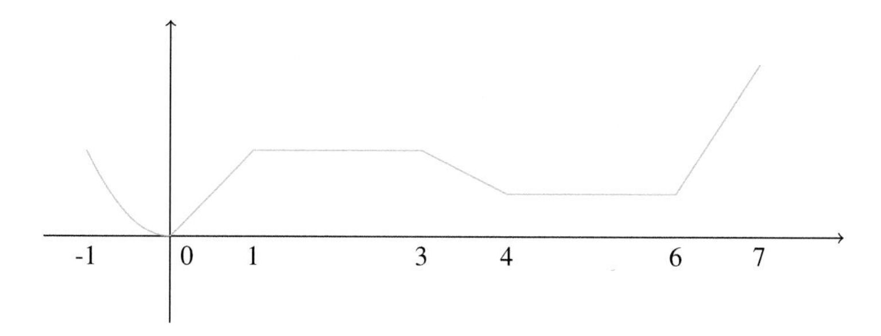

Figure 11.2 An absolutely continuous function that is not differentiable (see Exercise 11.4).

Exercise 11.5. Check to see that (11.7) defines a scalar product in $H_0^1(a,b]$.

Exercise 11.6.

(a) In the space $L^2(0,\infty)$ of (classes of) square integrable functions on the right half-axis we define the sequence $(x_n)_{n\geq 1}$ by $x_n = n1_{(n,n+1)}$ so that x_n equals n on the interval $(n, n+1)$ and 0 otherwise. Show that for each $s > 0$ the limit of $\lim_{n\to\infty} x_n(s)$ exists and equals 0 and yet $\|x_n\|_{L^2(0,\infty)} = n$, proving that the limit of $(x_n)_{n\geq 1}$ does not exist in $L^2(0,\infty)$.

(b) We define the following sequence of elements of $L^2[0,1]$. The first function equals 1 on the entire interval. The second equals one in $[0,\frac{1}{2}]$, and is zero otherwise, whereas the third equals 1 in $[\frac{1}{2},1]$ and is zero otherwise. The next four functions have value 1, respectively, on the first, second, third and fourth intervals of length $\frac{1}{4}$ that combined give $[0,1]$ – and are zero otherwise. The next eight functions are defined similarly, being equal to 1 on intervals of length $\frac{1}{8}$, etcetera. Check to see that these functions converge to 0 in $L^2[0,1]$, and yet that the limit $\lim_{n\to\infty} x_n(s)$ does not exist for any $s \in [0,1]$.

Exercise 11.7. Show that for v defined in (11.13), condition (d) (of the definition of solution to equation 11.12) is satisfied. **Hint:** Argue as in the proof of conditions (b) and (c), or use the Lebesgue dominated convergence theorem.

☞ CHAPTER SUMMARY

The fact that $L^2[0,\pi]$ is quite the same as ℓ^2 provides a way to solve the heat equation, one of the fundamental equations of mathematical physics (and of the theory of stochastic processes). As originally posed in $L^2[0,\pi]$, the equation seems to be rather difficult. But the isomorphism between $L^2[0,\pi]$ and ℓ^2 transforms it into a series of ordinary differential equations with constant coefficients, and these can be solved explicitly. On the level of calculations, we are simply using the well-known method of separation of variables of the theory of partial differential equations; more intrinsically, however, we are looking at the method from a proper perspective, the perspective of Hilbert spaces.

12

Completeness of the Space of Operators

In the previous chapters we have already encountered operators, that is, maps that assign functions to functions, or, more generally, vectors to vectors – these vectors are usually elements of normed spaces or Banach spaces. For example, the maps T of (3.17) and (3.22), which we have considered while studying Picard's theorem, are operators. The transformation studied in Chapter 5, devoted to the McKendrick–von Foerster equation, that is, the transformation I defined in (5.8), showing that the spaces $L^1(\mathbb{R}^+)$ and $L^1_\omega(\mathbb{R}^+)$ are indistinguishable (isometrically isomorphic), is also an operator. Operators are furthermore found in equations (5.10) and (6.2) – in this case we assign a vector from a Banach space to a function with values in this space. An important class of operators, the projection operators, was studied in Chapter 9 – even though we did not use the term. In Chapters 10 and 11 we have encountered the isometric isomorphism I, showing that the spaces $L^2[0,\pi]$ and ℓ^2 are indistinguishable. Furthermore, we have seen two copies of the same operator in these spaces (i.e., the operators A and B). A quest for other operators would reveal one also in Section 8.1 (namely the one called L).

In this chapter, we commence a systematic study of operators. The main theorem of this chapter says that the space of bounded linear operators mapping a normed linear space $\mathbb{X}$ into a Banach space $\mathbb{Y}$ is itself a Banach space. This elementary, but strikingly elegant, result allows treating operators as elements of another space; of course, the latter space differs from the original space $\mathbb{X}$, but has analogous structure – this shows them in a new light. In fact, we have encountered this fact already, at least a glimpse of it: $m \times n$ matrices are sometimes seen as transformations of $\mathbb{R}^n$ into $\mathbb{R}^m$, and sometimes as vectors in the space of matrices.

However, let's start from scratch.

12.1 Linear Operators

Let $\mathbb{X}$ and $\mathbb{Y}$ be linear spaces. An operator A mapping its domain $\mathcal{D}(A) \subset \mathbb{X}$ into $\mathbb{Y}$ is termed linear if the following conditions are met. First of all,

$$x, y \in \mathcal{D}(A), \alpha, \beta \in \mathbb{R} \Rightarrow \alpha x + \beta y \in \mathcal{D}(A);$$

second,

$$A(\alpha x + \beta y) = \alpha Ax + \beta Ay.$$

The first of these simply says that $\mathcal{D}(A)$ is a linear subspace of $\mathbb{X}$; it is the second that is key. We note that, in agreement with common custom, for linear operators, the parentheses around the argument are omitted in the notation: instead of $A(x)$ we simply write Ax (unless it leads to misunderstandings).

Let's have a look at some typical examples.

12.1 Example Multiplying a vector by a matrix. Let $\mathbb{X} = \mathbb{R}^n$ and $\mathbb{Y} = \mathbb{R}^m$, and let

$$A = \begin{pmatrix} a_{11} & \dots & a_{1n} \\ \dots & \dots & \dots \\ a_{m1} & \dots & a_{mn} \end{pmatrix}$$

be a matrix with m rows and n columns. The map

$$\mathbb{X} \ni x \mapsto Ax = A \begin{pmatrix} \xi_1 \\ \vdots \\ \xi_n \end{pmatrix} = \begin{pmatrix} a_{11} & \dots & a_{1n} \\ \dots & \dots & \dots \\ a_{m1} & \dots & a_{mn} \end{pmatrix} \begin{pmatrix} \xi_1 \\ \vdots \\ \xi_n \end{pmatrix} \in \mathbb{R}^m$$

is, as it is easy to see, linear. In this example, $\mathcal{D}(A)$ coincides with the entire $\mathbb{X}$.

12.2 Example Let $\mathbb{X}$ be the space of polynomials of degree at most n, and let $\mathbb{Y}$ be the space of polynomials of degree at most $n+1$. Given a $c \in \mathbb{R}$, to a polynomial

$$x(t) = a_n t^n + a_{n-1} t^{n-1} + \dots + a_1 t + a_0$$

we assign the polynomial

$$(Ax)(t) = a_n \frac{t^{n+1}}{n+1} + a_{n-1} \frac{t^n}{n} + \dots + a_1 \frac{t^2}{2} + a_0 t + c.$$

This map is linear if and only if the constant c is zero. We note that $\mathbb{X}$ may be identified with $\mathbb{R}^{n+1}$ since each polynomial may be identified with the vector of its coefficients, and similarly $\mathbb{Y}$ may be identified with $\mathbb{R}^{n+2}$. If we keep this identification in mind, this map (with $c = 0$) turns out to be a special case of

Example 12.1. Here again, $\mathcal{D}(A) = \mathbb{X}$. It is instructive to find an explicit form of the corresponding matrix.

12.3 Example Let both $\mathbb{X}$ and $\mathbb{Y}$ be the space $C_0(\mathbb{R})$ of continuous functions on $\mathbb{R}$ that have limits at both $+\infty$ and $-\infty$ equal to 0. The operator A that transforms a function x into y defined by

$$y(t) = x(t+1) + x(t-1), \qquad t \in \mathbb{R}$$

is linear. The reader should check that A maps $C_0(\mathbb{R})$ into $C_0(\mathbb{R})$, which, in particular, requires checking that the limits of y in plus and minus infinity are zero.

12.4 Example Let $P = (p_{i,j})_{i,j\geq 1}$ be an infinite *stochastic matrix*. That is, we assume that $p_{ij} \geq 0$ and, for each $i \geq 1$, $\sum_{j=1}^{\infty} p_{ij} = 1$. Furthermore, let $\mathbb{X} = \mathbb{Y} = \ell^1$ where, as before, ℓ^1 is the space of absolutely summable sequences $(\xi)_{i\geq 1}$. A map (denoted by the same letter as P, as it may be identified with that matrix) that to a vector $(\xi)_{i\geq 1} \in \ell^1$ assigns the vector

$$Px = y = (\xi_1, \xi_2, \dots) \begin{pmatrix} p_{11} & p_{11} & \cdots \\ p_{21} & p_{22} & \cdots \\ \cdots & \cdots & \cdots \end{pmatrix}, \tag{12.1}$$

is linear. All is clear here, except perhaps for the fact that y belongs to ℓ^1 if x does. This is, however, proved by the following calculation: letting $\eta_i = \sum_{j=1}^{\infty} \xi_j p_{ji}$ be the ith coordinate of $Px = y$, we have

$$\begin{aligned} \|Px\| = \sum_{i=1}^{\infty} |\eta_i| &\leq \sum_{i=1}^{\infty}\sum_{j=1}^{\infty} |\xi_j p_{ji}| = \sum_{i=1}^{\infty}\sum_{j=1}^{\infty} |\xi_j| p_{ji} \\ &= \sum_{j=1}^{\infty} |\xi_j| \sum_{i=1}^{\infty} p_{ji} = \sum_{j=1}^{\infty} |\xi_j| = \|x\| < \infty. \end{aligned} \tag{12.2}$$

(The last inequality justifies changing the order of summation in the third step.) By the way, if you are familiar with Markov chains, the operator we are discussing here has a nice interpretation in stochastic terms. Namely, P may be interpreted as a transition probability matrix for a Markov chain. Then, if x has non-negative coordinates summing to one, so that x may be interpreted as an initial distribution of the chain, then Px is the distribution of the chain after one step.

12.5 Example Let c_{00} be the space of sequences $x = (\xi_i)_{i\ge 1}$ which have all but a finite number of non-zero coordinates. The operator A defined by

$$Ax = \sum_{i=1}^{\infty} i\xi_i$$

is linear and maps c_{00} into $\mathbb{R}$. We note that the defining sum has in fact only a finite number of non-zero terms, and thus is well defined. Operators that have scalar values are termed *functionals*.

12.6 Example Let $\mathbb{X} = C^1[0,1]$ be the space of real functions on $[0,1]$ that are continuously differentiable, and let $\mathbb{Y} = C[0,1]$ be the space of continuous functions on this interval. The operator that assigns the derivative $x' \in C[0,1]$ to an $x \in C^1[0,1]$ is linear.

12.7 Example In the space c_0 of sequences $x = (\xi_i)_{i\ge 1}$ that converge to zero we define two operators:

$$Ax = (\xi_2, \xi_3, \ldots)$$

and

$$Bx = (0, \xi_1, \xi_2, \ldots).$$

These are known as the shift to the left and the shift to the right, respectively, and are linear.

12.8 Example Let $\mathbb{X} = \mathbb{Y} = C_0[0,\infty)$ be the space of continuous real functions x on $\mathbb{R}^+$, such that $\lim_{t\to 0} x(t)$ exists and equals 0. Also, let X be an exponentially distributed random variable with parameter λ. This means that the probability density function of X is $s \mapsto \lambda \mathrm{e}^{-\lambda s}, s \ge 0$. We define an operator A by

$$Ax(t) = E\,x(t+X) = \lambda \int_0^{\infty} \mathrm{e}^{-\lambda s} x(t+s)\,\mathrm{d}s, \qquad t \ge 0,$$

where E denotes expected value. Then, again, A is linear.

12.9 Example Suppose that A and B with the same domain have values in the same linear space $\mathbb{Y}$. If α and β are real numbers, and A and B are linear, then so is the operator $\alpha A + \beta B$ defined by $(\alpha A + \beta B)x = \alpha Ax + \beta Bx$.

12.10 Example Let $\mathbb{X} = \mathbb{Y}$. The operator that maps each $x \in \mathbb{X}$ into itself is called the identity operator and is obviously linear. This operator is denoted $I_{\mathbb{X}}$ or simply I.

12.11 Example Any $y \in C[a,b]$ provides an example of a linear functional, as follows. To an $x \in C[a,b]$ we assign the number $Ax := \int_a^b y(t)x(t)\,dt$. It is easy to check that $A\colon C[a,b] \to \mathbb{R}$ is linear.

12.12 Example Much as in the previous example, any y in a Hilbert space $\mathbb{H}$ allows us to define a functional in this space by $Ax = (x, y)$. In contrast to the previous example, though, in a Hilbert space all (bounded) linear functionals are of this form (see Section 12.4), but in $C[a,b]$ this is not the case.

The reader should check which of the operators recalled in the introduction to this chapter are linear (Exercise 12.4).

12.2 Bounded Operators and Their Norms

12.2.1 Bounded Operators

A linear map A from a normed linear space $\mathbb{X}$ to a normed linear space $\mathbb{Y}$ is said to be bounded if there is a constant $M \geq 0$ such that for all $x \in \mathbb{X}$,

$$\|Ax\|_{\mathbb{Y}} \leq M\|x\|_{\mathbb{X}};$$

in particular, we assume that the domain of A is $\mathbb{X}$.

Let's look at the operators of Section 12.1. In Example 12.1 assume, to fix attention, that $\|x\|_{\mathbb{X}} = \max_{i=1,\dots,n} |\xi_i|$ and $\|y\|_{\mathbb{Y}} = \sqrt{\sum_{i=1}^{m} \eta_i^2}$. Then, introducing $M_0 := \max_{i,j} |a_{i,j}|$, we obtain

$$\|Ax\| = \sqrt{\sum_{i=1}^{m} \eta_i^2} = \sqrt{\sum_{i=1}^{m}\left(\sum_{j=1}^{n} a_{i,j}\xi_j\right)^2} \leq \sqrt{\sum_{i=1}^{m}\left(M_0\sum_{j=1}^{n} |\xi_j|\right)^2}$$
$$\leq \sqrt{mM_0^2 n^2\|x\|^2} = \sqrt{m}M_0 n\|x\|.$$

This shows that A is bounded, and that for M from the definition one can take any number no smaller than $\sqrt{m}M_0 n$.

Turning to Example 12.2 we note the following, alternative formula for A:

$$Ax(t) = \int_0^t x(s)\,ds.$$

If both $\mathbb{X}$ and $\mathbb{Y}$ are equipped with the maximum norm (that is, in both cases $\|x\| = \max_{t\in[0,1]} |x(t)|$), we have the following estimates:

$$\|Ax\| = \max_{t\in[0,1]} |Ax(t)| = \max_{t\in[0,1]} \left| \int_0^t x(s)\,\mathrm{d}s \right| \le \max_{t\in[0,1]} \int_0^t |x(s)|\,\mathrm{d}s$$
$$\le \max_{t\in[0,1]} \int_0^t \|x\|\,\mathrm{d}s = \|x\|,$$

proving that A is bounded; any constant M larger or equal to 1 will do.

We skip Example 12.3 for the time being; we will cover it in detail later. As to 12.4, we note that (12.2) proves boundedness of the operator from this example (again, $M = 1$).

The operator of Example 12.5, in turn, is not bounded; we assume here that the norm in c_{00} is defined by $\|x\| = \max_{i\ge 1} |\xi_i|$. To show this, we consider e_n of (10.1) as a member of c_{00}. We have $\|e_n\| = 1$, whereas $Ae_n = n$. Therefore, if constant M from the definition exits, we have $M \ge n$. This, however, is impossible, because n is arbitrary.

There is no simple answer to the question of whether the operator of Example 12.6 is bounded or not, because the answer depends on the norms we choose. Certainly, this is true of any operator in any space, for boundedness obviously depends on the choice of norms involved: for one norm the operator will be bounded; for another it might not. Here, we would like to choose a 'natural' norm. For instance, if both spaces are simply equipped with the supremum norm, the derivative is not a bounded operator (and $C^1[0,1]$ is not a Banach space). Indeed, let $x_n(t) := t^n, t \in [0,1]$. Then $\|x\|_{C[0,1]} = 1$, whereas $x_n'(t) = nt^{n-1}, t \in [0,1]$, implying $\|x_n'\| = n$. Thus, as in the previous example, the M of the definition cannot exist.

However, if $C^1[0,1]$ is equipped with

$$\|x\|_{C^1[0,1]} = \|x\|_{C[0,1]} + \|x'\|_{C[0,1]},$$

the situation changes to our benefit: the space is complete and the operator is bounded. For, obviously,

$$\|x'\|_{C[0,1]} \le \|x\|_{C^1[0,1]}.$$

12.2.2 The Space of Bounded Operators

Leaving aside temporarily the question of boundedness of the remaining operators of Section 12.1, we turn to a description of the space of bounded operators in the general context. We begin by noting that bounded linear operators that map one normed linear space, say $\mathbb{X}$, into another normed linear space, form a linear space themselves. This space is denoted

$$\mathcal{L}(\mathbb{X}, \mathbb{Y}).$$

More precisely, if A and B belong to this space and α and β are scalars, then $\alpha A + \beta B$ can be defined by the formula

$$(\alpha A + \beta B)(x) = \alpha Ax + \beta Bx, \qquad x \in \mathbb{X},$$

and this definition makes $\mathcal{L}(\mathbb{X}, \mathbb{Y})$ a linear space. In particular, we note that the combined conditions $\|Ax\| \le L\|x\|$ and $\|Bx\| \le M\|x\|$ imply

$$\|(\alpha A + \beta B)(x)\| \le (|\alpha|L + |\beta|M)\|x\|,$$

and this shows that the operator $\alpha A+\beta B$ is bounded (for instance, the operator of Example 12.9 is bounded). That is, addition and multiplication by scalars does not lead out of $\mathcal{L}(\mathbb{X}, \mathbb{Y})$.

In the case where $\mathbb{X} = \mathbb{Y}$, instead of $\mathcal{L}(\mathbb{X}, \mathbb{Y})$ we write

$$\mathcal{L}(\mathbb{X})$$

and speak of the space of bounded linear operators in $\mathbb{X}$. If $\mathbb{Y} = \mathbb{R}$, the notation is even simpler: $\mathbb{X}^*$; this space is known as the space of bounded linear functionals on $\mathbb{X}$.

The following theorem explains why, in the case of linear operators, the adjectives *bounded* and *continuous* are used interchangeably.

12.13 Theorem *Let $\mathbb{X}$ and $\mathbb{Y}$ be linear normed spaces, and let $A \colon \mathbb{X} \to \mathbb{Y}$ be a linear operator. The following conditions are equivalent.*

(a) A is continuous.
(b) A is continuous at a certain $x \in \mathbb{X}$.
(c) A is continuous at 0.
(d) $\sup_{\|x\|_{\mathbb{X}}=1} \|Ax\|_{\mathbb{Y}}$ is finite.
(e) A is bounded.

If A is continuous,

$$\sup_{\|x\|_{\mathbb{X}}=1} \|Ax\|_{\mathbb{Y}} = \min\{M \in \mathcal{S}\}, \tag{12.3}$$

where $\mathcal{S}$ is the set of positive numbers such that $\|Ax\|_{\mathbb{Y}} \le M\|x\|_{\mathbb{X}}$ for all $x \in \mathbb{X}$.

Proof Implication $(a) \Rightarrow (b)$ is obvious, because (a) says, by definition, that for any $x \in \mathbb{X}$ the sequence $(Ax_n)_{n\ge1}$ converges to Ax as long as $(x_n)_{n\ge1}$ converges to x, and (b) says that this is the case only for a certain x.

Next, assume (b) and let x be the vector at which A is continuous. Suppose that $\lim_{n\to\infty} x_n = 0$. Then $\lim_{n\to\infty}(x_n + x) = x$. Hence, since (b) is true, the sequence $(A(x_n + x))_{n\geq 1} = (Ax_n + Ax)_{n\geq 1}$ converges to Ax; in other words, $\lim_{n\to\infty} Ax_n = 0$, proving (c).

For the proof that (c) implies (d), we assume that (d) does not hold, that is, there are $x_n \in \mathbb{X}$ such that $\|x_n\|_{\mathbb{X}} = 1$ and yet $\|Ax_n\|_{\mathbb{Y}} > n$. In such a case, the vectors $y_n = \frac{1}{\sqrt{n}}x_n$ form a sequence converging to 0, whereas $\|Ay_n\|_{\mathbb{Y}} > \sqrt{n}$, showing that $(Ax_n)_{n\geq 1}$ cannot converge to anything, and to zero in particular (it is easy to see that one always has $A0 = 0$). Since this contradicts (c), we are done.

To check that (d) implies (e) we define $M := \sup_{\|x\|_{\mathbb{X}}=1} \|Ax\|_{\mathbb{Y}}$, and claim that

$$\|Ax\|_{\mathbb{Y}} \leq M\|x\|_{\mathbb{X}}.$$

Indeed, this inequality is obvious for $x = 0$; also, for a non-zero x, the norm of $\frac{1}{\|x\|_{\mathbb{X}}}x$ is 1, implying $\|A\frac{1}{\|x\|_{\mathbb{X}}}x\|_{\mathbb{Y}} \leq M$. It remains to multiply both sides of this relation by $\|x\|_{\mathbb{X}}$ to obtain the claim in this case.

Finally, we note that (a) is a consequence of (e) since $\|Ax_n - Ax\|_{\mathbb{Y}} \leq \|A(x_n - x)\| \leq M\|x_n - x\|_{\mathbb{X}}$.

To prove the second part of our theorem, we note that, on the one hand, in the proof of implication (d) $\Rightarrow$ (e) we have shown that $M_1 := \sup_{\|x\|_{\mathbb{X}}=1} \|Ax\|_{\mathbb{Y}}$ belongs to $\mathcal{S}$. On the other hand, if $\|Ax\|_{\mathbb{Y}} \leq M\|x\|_{\mathbb{X}}$ holds for all $x \in X$, then by restricting ourselves to x of norm 1 we see that $M_1 \leq M$. This shows, however, that M_1 is the minimal element of the set $\mathcal{S}$. □

12.2.3 Operator Norm

The second part of Theorem 12.13 is worth closer attention. If A is bounded, the set $\mathcal{S}$ is non-empty. There are then infinitely many constants M that belong to $\mathcal{S}$ – together with M, this set contains all the numbers that are larger than it. The quantity

$$\|A\| := \sup_{\|x\|_{\mathbb{X}}=1} \|Ax\|_{\mathbb{Y}} = \min\{M \in \mathcal{S}\} \tag{12.4}$$

introduced in (12.3) is the smallest of all elements of $\mathcal{S}$, and it is calculated effectively. This quantity is termed the norm of the operator A. The reader should check that $A \mapsto \|A\|$ satisfies all the conditions of norm, as defined in Section 4.1.2.

It is worth summarizing our analysis as follows.

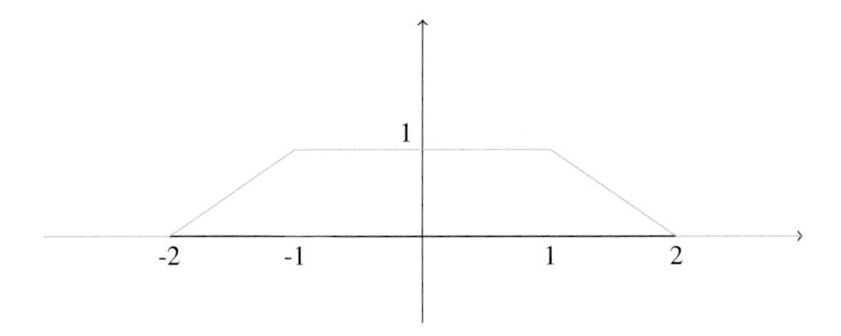

Figure 12.1 A function used in the analysis of Example 12.3.

> The space of bounded linear operators from one linear normed space to another linear normed space is itself normed.

Let's calculate some norms explicitly, beginning with Example 12.3. The operator A described there is bounded, because

$$\|Ax\| = \sup_{t\in\mathbb{R}} |y(t)| \le \sup_{t\in\mathbb{R}} |x(t+1)| + \sup_{t\in\mathbb{R}} |x(t-1)| = 2\sup_{t\in\mathbb{R}} |x(t)| = 2\|x\|.$$

These inequalities show also that $\|A\| \le 2$. To convince ourselves that $\|A\| = 2$ we consider the function depicted in Figure 12.1. Obviously, $\|x\| = 1$, whereas $Ax(0) = x(1) + x(-1) = 2$. Thus $\|Ax\| \ge |Ax(0)| \ge 2$, and so $\|A\| \ge 2$, completing the proof.

Next, we show that all operators related to stochastic matrices, as in Example 12.4, have norm one. A look at (12.2) convinces us that $\|P\| \le 1$. However, if x is a vector with non-negative coordinates, all inequalities in 12.2 can be replaced by equalities, rendering $\|Px\| = \|x\|$. In particular, for non-negative x with norm 1 (i.e., for x with non-negative coordinates that sum up to 1), $\|Px\| = 1$. It follows that $\sup_{\|x\|=1} \|Px\| = 1$, that is, $\|P\| = 1$, as claimed. For future use, we note that the operators related to stochastic matrices are termed *Markov operators*. Moreover, operators with norms equal to 1 are said to be *contractions*.[1]

More generally, a matrix $A = \left(a_{i,j}\right)_{i,j\in\mathbb{N}}$ defines a bounded operator in l^1 given by

[1] The latter definition seems to be confusing, as it does not agree with the notion of contraction used in the context of Banach's principle, but in practice this established terminology works well.

$$A(\xi_i)_{i\geq 1} = \left(\sum_{j=1}^{\infty} \xi_j a_{j,i}\right)_{i\geq 1}, \tag{12.5}$$

as long as $\sup_{j\in\mathbb{N}} \sum_{i\in\mathbb{N}} |a_{j,i}|$ is finite, and then

$$\|A\| = \sup_{j\in\mathbb{N}} \sum_{i=1}^{\infty} |a_{j,i}|. \tag{12.6}$$

(Here, as above, the operator is identified with the matrix, and thus denoted by the same symbol.) In words, $A(\xi_i)_{i\geq 1}$ is obtained by multiplying a row vector $(\xi_i)_{i\geq 1}$ by the matrix A (cf. (12.1)).

It is almost immediate that the norm of A does not exceed the expression on the right of (12.6), let's call it M_0, because

$$\left\| \left(\sum_{j=1}^{\infty} \xi_j a_{j,i}\right)_{i\geq 1} \right\| = \sum_{i=1}^{\infty} \left| \sum_{j=1}^{\infty} \xi_j a_{j,i} \right| \leq \sum_{i=1}^{\infty} \sum_{j=1}^{\infty} |\xi_j|\,|a_{j,i}| = \sum_{j=1}^{\infty} |\xi_j| \sum_{i=1}^{\infty} |a_{j,i}|$$
$$\leq M_0 \sum_{j=1}^{\infty} |\xi_j| = M_0 \left\|(\xi_i)_{i\geq 1}\right\| < \infty,$$

and again changing the order of summation in the intermediate step is justified by the final inequality.

To show that $\|A\| \geq M_0$, we consider e_n of (10.1) as a member of ℓ^1. Then $\|e_n\| = 1$ and $Ae_n = \left(a_{n,i}\right)_{i\geq 1}$ (the right-hand side here is the nth row of matrix A), showing that $\|Ae_n\| = \sum_{i=1}^{\infty} |a_{n,i}|$. Thus, for each n, $\|A\| \geq \|Ae_n\| = \sum_{i=1}^{\infty} |a_{n,i}|$. Therefore, A cannot be bounded if M_0 is infinity, and if M_0 is finite we have $\|A\| \geq M_0$. The reader should check to see that this result contains the statement that Markov operators have norm one as a special case.

Both operators in Example 12.7 are bounded and their norms are 1. For, on the one hand,

$$\|Ax\| = \sum_{i=2}^{\infty} |\xi_i| \leq \sum_{i=1}^{\infty} |\xi_i| = \|x\|,$$

and, on the other hand, if $\xi_1 = 0$ the inequality here may be replaced by equality. With B we proceed similarly.

In Example 12.12 the operator is bounded with norm $\|y\|$. Indeed, by the Cauchy–Schwarz inequality, $|Ax| = |(x,y)| \leq \|x\|\,\|y\|$ and, on the other hand, for $x = \frac{y}{\|y\|}$ (the case of $y = 0$ is trivial), we have $\|x\| = 1$ and $Ax = \|y\|$.

Examples 12.8 and 12.11 are perhaps the most instructive. Here, we will compute the norm even though we will not be able to find an $x \in \mathbb{X}$ of norm

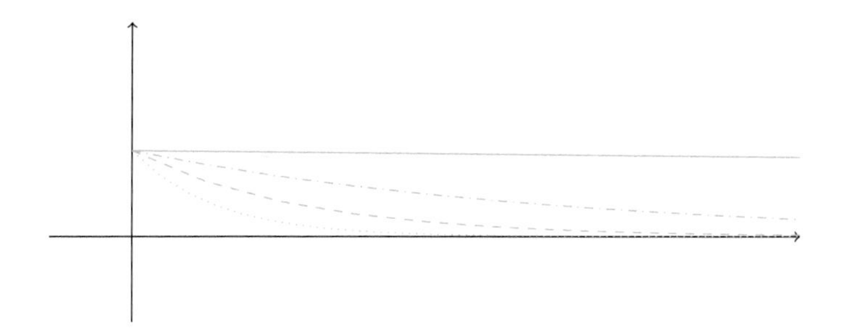

Figure 12.2 Exponential functions $e_\mu(t) := \mathrm{e}^{-\mu t}, t \geq 0$ for small μ (dotted: $\mu = 1$, dash-dotted: $\mu = 0.5$, dashed: $\mu = 0.2$ and solid: $\mu = 0.01$). As $\mu \to 0$ these functions converge in a sense to e_0, which equals 1 for all $t \geq 0$.

one such that $\|Ax\| = \|A\|$. In the first case, that is, in Example 12.8, it is easy to check that $\|A\| \leq 1$ using $\lambda \int_0^\infty \mathrm{e}^{-\lambda s}\, \mathrm{d}s = 1$:

$$\begin{aligned}\|Ax\| &= \sup_{t\geq 0} |\lambda \int_0^\infty \mathrm{e}^{-\lambda s} x(t+s)\, \mathrm{d}s| \leq \sup_{t\geq 0} \lambda \int_0^\infty \mathrm{e}^{-\lambda s} |x(t+s)|\, \mathrm{d}s \\ &\leq \sup_{t\geq 0} \lambda \int_0^\infty \mathrm{e}^{-\lambda s} \|x\|\, \mathrm{d}s = \|x\|.\end{aligned}$$

If the function e_0, which, by definition, equals 1 for all $t \geq 0$, were a member of $C_0[0,\infty)$, the proof of $\|A\| \geq 1$ would have been immediate; for, Ae_0 would have been e_0 by definition. But it is not, since $\lim_{t\to\infty} e_0(t) = 1 \neq 0$. Therefore, we need to be more careful. To this end, for each $\mu > 0$, we define $e_\mu \in C_0[0,\infty)$ by $e_\mu(t) = \mathrm{e}^{-\mu t}, t \geq 0$, see Figure 12.2. Clearly $\|e_\mu\| = e_\mu(0) = 1$, whereas

$$Ae_\mu(t) = \lambda \int_0^\infty \mathrm{e}^{-\lambda s} \mathrm{e}^{-\mu(t+s)}\, \mathrm{d}s = \lambda \mathrm{e}^{-\mu t} \int_0^\infty \mathrm{e}^{-(\lambda+\mu)s}\, \mathrm{d}s = \frac{\lambda}{\lambda+\mu} \mathrm{e}^{-\mu t}, \quad t \geq 0.$$

Therefore, $\|Ae_\mu\| = \frac{\lambda}{\lambda+\mu}$, and so

$$\|A\| \geq \sup_{\mu>0} \|Ae_\mu\| \geq \lim_{\mu\to 0} \|Ae_\mu\| = 1.$$

In Example 12.11, we claim that A is bounded with norm equaling $\int_a^b |y(t)|\, \mathrm{d}t$. To prove this fact, we note first that (12.4) may also be written as

$$\|A\| := \sup_{\|x\|_{\mathbb{X}} \leq 1} \|Ax\|_{\mathbb{Y}}. \tag{12.7}$$

Indeed, by taking the supremum over a larger set we apparently increase $\|A\|$, but in fact we do not, because if $0 < \|x\| < 1$ then for $x_0 = \frac{x}{\|x\|}$ we have

$\|x_0\| = 1$ and $\|Ax_0\| = \frac{1}{\|x\|}\|Ax\| > \|Ax\|$. Notably, being able to work with x's that have norm at most one, and not only those that have norm one, comes in handy, as we will now see.

First of all, though, the obvious estimate

$$|Ax| \leq \int_a^b |y(t)|\,\mathrm{d}t\,\|x\|$$

shows that $\|A\| \leq \int_a^b |y(t)|\,\mathrm{d}t$. To prove that $\|A\|$ in fact is equal to $\int_a^b |y(t)|\,\mathrm{d}t$ we define $x_n(t) := \frac{ny(t)}{1+n|y(t)|}, t \in [a,b], n \geq 1$. Then $\|x_n\| \leq 1$ (but in general $\|x_n\| \neq 1$). Since

$$0 \leq \int_a^b |y(t)|\,\mathrm{d}t - Ax_n = \frac{1}{n}\int_a^b |x_n(t)|\,\mathrm{d}t \leq \frac{1}{n}(b-a)\|x_n\| \leq \frac{b-a}{n},$$

we have $\lim_{n\to\infty} Ax_n = \int_a^b |y(t)|\,\mathrm{d}t$. This shows our claim.

12.3 Completeness of the Space of Operators

Since the space $\mathcal{L}(\mathbb{X},\mathbb{Y})$ of operators from a normed space to a normed space is itself normed, we have a natural notion of convergence of operators: a sequence $(A_n)_{n\geq 1}$ converges to A if

$$\lim_{n\to\infty} \|A_n - A\| = 0.$$

More precisely: in this case, we say that $(A_n)_{n\geq 1}$ converges to A in the operator norm (or simply: in the norm).

For instance, let $\mathbb{X}$ be a normed linear space. By inequality (12.10), which the reader will prove in Exercise 12.3, if a sequence $(A_n)_{n\geq 1}$ of bounded linear operators in $\mathbb{X}$ converges to an A in the same space, then for any operator $B \in \mathcal{L}(\mathbb{X})$, the sequences $(BA_n)_{n\geq 1}$ and $(A_nB)_{n\geq 1}$ converge to BA and AB, respectively. Similarly, let $\mathbb{X}, \mathbb{Y}$ and $\mathbb{Z}$ be normed linear spaces. If a sequence $(A_n)_{n\geq 1}$ of operators in $\mathcal{L}(\mathbb{X},\mathbb{Y})$ converges to an A and a sequence $(B_n)_{n\geq 1}$ of operators in $\mathcal{L}(\mathbb{Y},\mathbb{Z})$ converges to a B, then $(B_nA_n)_{n\geq 1}$ converges to BA. To see this, we write

$$\|B_nA_n - BA\| \leq \|B_nA_n - B_nA\| + \|B_nA - BA\| \leq \|B_n\|\,\|A_n - A\| + \|A\|\,\|B_n - B\|$$

and use the fact that the numerical sequence $(\|B_n\|)_{n\geq 1}$ converges to $\|B\|$ (cf. Exercise 1.2).

It is natural to ask whether $\mathcal{L}(\mathbb{X},\mathbb{Y})$ is a complete space. The answer is in the affirmative provided $\mathbb{Y}$ is complete.

12.14 Theorem *If $\mathbb{Y}$ is a Banach space, so is $\mathcal{L}(\mathbb{X}, \mathbb{Y})$.*

Proof Suppose that $(A_n)_{n\geq 1}$ is a fundamental sequence in $\mathcal{L}(\mathbb{X}, \mathbb{Y})$. Then, for all $x \in \mathbb{X}$ and $\epsilon > 0$,

$$\|A_n x - A_m x\| \leq \|A_n - A_m\| \, \|x\| \leq \epsilon \|x\|, \tag{12.8}$$

as long as n and m are are sufficiently large. This shows that $(A_n x)_{n\geq 1}$ is a Cauchy sequence in $\mathbb{Y}$. Completeness of $\mathbb{Y}$ implies then that $(A_n x)_{n\geq 1}$ has a limit. Since this limit obviously depends on x, we obtain a map $x \mapsto Ax := \lim_{n\to\infty} A_n x$. There is no doubt about the condition $A(\alpha x + \beta y) = \alpha Ax + \beta Ay$ (where $\alpha, \beta \in \mathbb{R}$ and $x, y \in \mathbb{X}$), proving that A is linear. Letting $m \to \infty$ in (12.8), we see that for all sufficiently large n and all $x \in \mathbb{X}$ (with n independent of x),

$$\|A_n x - Ax\| \leq \epsilon \|x\|.$$

This shows two things. First of all, A is bounded, because $\|Ax\| \leq \|Ax - A_n x\| + \|A_n x\| \leq (\epsilon + \|A_n\|)\|x\|$, that is, A belongs to $\mathcal{L}(\mathbb{X}, \mathbb{Y})$. Second, $(A_n)_{n\geq 1}$ converges to A, the inequality above being equivalent to $\|A_n - A\| \leq \epsilon$ (for sufficiently large n). □

12.4 General Form of Bounded Linear Functionals in Hilbert Spaces

In the part of Section 12.2.3 devoted to Example 12.11 we have seen that for any element y of a Hilbert space $\mathbb{H}$,

$$\mathbb{H} \ni x \mapsto (x, y)$$

is a bounded linear functional with norm equal to $\|y\|$. Here, we will show the converse: any bounded linear functional on $\mathbb{H}$ is of this form. This result is known as the Riesz representation theorem.

To see this, given a bounded functional $A \colon \mathbb{H} \to \mathbb{R}$, think of the subspace, say $\mathbb{H}_0$, of x such that $Ax = 0$, that is, about the kernel of A. Since A is assumed bounded, $\mathbb{H}_0$ is closed. If $\mathbb{H}_0$ is the entire $\mathbb{H}$, which is the case where $Ax = 0$ for all x, we have $Ax = (x, y)$ for $y = 0$.

Assume therefore that $\mathbb{H} \setminus \mathbb{H}_0$ is non-empty, and take a $z_0 \in \mathbb{H} \setminus \mathbb{H}_0$. Then $z_0 - Pz_0$, where Pz_0 is the projection of z_0 on $\mathbb{H}_0$, is non-zero, because z_0 does not belong to $\mathbb{H}_0$ but Pz_0 does. By dividing $z_0 - Pz_0$ by its length, we obtain a vector of norm one: $z := \frac{z_0 - Pz_0}{\|z_0 - Pz_0\|}$. Finally, we define $y := (Az)z$. This vector is non-zero (because neither z nor Az is zero) and perpendicular to all elements of $\mathbb{H}_0$. Moreover, since Ay and (y, y) are equal to $(Az)^2$, we have $Ay = \|y\|^2$. It follows that for any $x \in \mathbb{H}$, $A(x - \frac{Ax}{\|y\|^2} y) = 0$. Hence, $x - \frac{Ax}{\|y\|^2} y$ is perpendicular to y, and so

$$(x, y) = \left(x - \frac{Ax}{\|y\|^2} y, y\right) + \left(\frac{Ax}{\|y\|^2} y, y\right) = Ax,$$

as claimed.

Let's summarize. In the Hilbert space case any element of the space $\mathbb{H}^*$ of functionals can be identified with an element of the original space $\mathbb{H}$, and this identification is complete: the norm of the functional in $\mathbb{H}^*$ is the same as the norm of the element of $\mathbb{H}$. In the case of Banach spaces a similar statement is not true (see Exercise 12.10).

For example, in ℓ^2 there are no (bounded and linear) functionals other than those of the form

$$\ell^2 \ni (\xi_i)_{i \geq 1} \mapsto \sum_{i=1}^{\infty} \xi_i \eta_i \in \mathbb{R},$$

where $(\eta_i)_{i \geq 1} \in \ell^2$. Moreover, the norm of the functional of this form coincides with $\| (\eta_i)_{i \geq 1} \|_{\ell^2}$. Similarly in $L^2[a, b]$: for any functional defined in this space there is a unique $y \in L^2[a, b]$ such that

$$Fx = \int_a^b x(s) y(s) \, \mathrm{d}s, \qquad x \in L^2[a, b],$$

and $\|F\|_{(L^2[a,b])^*} = \|y\|_{L^2[a,b]}$.

12.5 Exercises

Exercise 12.1. Here is an exercise in linear operations. Suppose $a_{i,j}, i = 1, \ldots, n,\ j = 1, \ldots, n-1$, where $n \geq 2$, are real numbers, and $x_1, \ldots, x_n$ are real integrable functions on $\mathbb{R}$. Check to see that

$$\int_{\mathbb{R}} \begin{vmatrix} a_{1,1} & a_{1,2} & \cdots & a_{1,n-1} & x_1(t) \\ a_{2,1} & a_{2,2} & \cdots & a_{2,n-1} & x_2(t) \\ \vdots & \vdots & \cdots & \ddots & \vdots \\ a_{n,1} & a_{n,2} & \cdots & a_{n,n-1} & x_n(t) \end{vmatrix} \mathrm{d}t = \begin{vmatrix} a_{1,1} & a_{1,2} & \cdots & a_{1,n-1} & \int_{\mathbb{R}} x_1(t)\mathrm{d}t \\ a_{2,1} & a_{2,2} & \cdots & a_{2,n-1} & \int_{\mathbb{R}} x_2(t)\mathrm{d}t \\ \vdots & \vdots & \cdots & \ddots & \vdots \\ a_{n,1} & a_{n,2} & \cdots & a_{n,n-1} & \int_{\mathbb{R}} x_n(t)\mathrm{d}t \end{vmatrix}. \tag{12.9}$$

Specialize to the case where, for a certain $\ell = 0, \ldots, n-2$ and a real function y on $\mathbb{R}$ such that

$$\int_{\mathbb{R}} |t^k y(t)| \, \mathrm{d}t < \infty, \qquad k = 0, \ldots, 2n-3,$$

we have

$$x_i(t) = t^{i+\ell-1} y(t),\ t \in \mathbb{R}, i = 1, \dots, n, \text{ whereas } a_{i,j} = \int_{\mathbb{R}} t^{i+j-2} y(t)\, \mathrm{d}t,$$

and then check that the integral on the left of (12.9) is zero. (The reader might have noticed that we are brushing against so-called Hankel determinants here: just take $\ell = n - 1$.)

Exercise 12.2. Calculate the norms of all bounded linear operators of Section 12.1 (some of them are already calculated in the text).

Exercise 12.3. Let $\mathbb{X}, \mathbb{Y}$ and $\mathbb{Z}$ be linear normed spaces, and let $A \in \mathcal{L}(\mathbb{X}, \mathbb{Y})$ and $B \in \mathcal{L}(\mathbb{Y}, \mathbb{Z})$ be bounded linear operators. The operator C, defined by $Cx = B(Ax), x \in \mathbb{X}$ and denoted BA, is said to be the composition of A and B. Show that BA is bounded also, that is, that $BA \in \mathcal{L}(\mathbb{X}, \mathbb{Z})$, and that

$$\|BA\| \le \|B\| \, \|A\|. \tag{12.10}$$

Exercise 12.4. Examine the operators mentioned in the introductory section to this chapter to see if they are linear and bounded. Calculate norms of those operators that are linear and bounded. In particular, note that for the map $x \mapsto Px$ described in Theorem 9.4 we have $P^2 = P$, and use Exercise 12.3 to conclude that $\|P\| \ge 1$; then argue to show that in fact $\|P\| = 1$.

Exercise 12.5. Let $\mathbb{X}$ be a Banach space. An operator $P \in \mathcal{L}(\mathbb{X})$ is said to be a projection if $P^2 = P$. Show that $Q := I - P$, where I is the identity operator, is a projection whenever P is.

Exercise 12.6. As we have seen in Exercise 12.4, projections in a Hilbert space have norm equal to 1. This is not the case in Banach spaces. For example, consider $\mathbb{R}^2$ with norm

$$\|(\xi_1, \xi_2)\| = \max(|\xi_1|, |\xi_2|), \qquad (\xi_1, \xi_2) \in \mathbb{R}^2,$$

and given $a > 0$ define the operator $P \colon \mathbb{R}^2 \to \mathbb{R}^2$ by

$$P(\xi_1, \xi_2) = (\xi_1 + a\xi_2, 0),$$

so that P projects $\mathbb{R}^2$ on the subspace $\mathbb{X}_0 \subset \mathbb{R}^2$ composed of $(\xi_1, \xi_2) \in \mathbb{R}^2$ such that $\xi_2 = 0$. Check to see that $P^2 = P$ and $\|P\| = 1 + a$. **Hint:** Think of $\xi_1 = \xi_2 = 1$.

Exercise 12.7. Let $x \colon [a, b] \mapsto \mathbb{X}$ be a continuous function with values in a Banach space $\mathbb{X}$, and let $A \in L(\mathbb{X}, \mathbb{Y})$ be a bounded linear operator mapping $\mathbb{X}$ into another Banach space $\mathbb{Y}$. Show that

$$A \int_a^b x(t)\,\mathrm{d}t = \int_a^b Ax(t)\,\mathrm{d}t. \tag{12.11}$$

Hint: Think of Ax_n, where x_n is an approximating sum for $\int_a^b x(t)\,\mathrm{d}t$.

Exercise 12.8. Let $C[0,1]$ be the space of real continuous functions on the unit interval, and let $C[0,\infty]$ be the space of real continuous functions on the right half-axis that have finite limits at $+\infty$. The formula

$$Ax(t) = x(\mathrm{e}^{-t}), \qquad t \geq 0$$

defines a bounded linear operator mapping $C[0,1]$ onto $C[0,\infty]$ such that $\|Ax\| = \|x\|$. Find the operator B mapping $C[0,\infty]$ onto $C[0,1]$ such that AB is the identity operator in $C[0,\infty]$ and BA is the identity operator in $C[0,1]$. Also, using the Weierstrass theorem (not the Stone–Weierstrass theorem), prove the result described in Exercise 7.5.

Exercise 12.9. Let $C[0,\infty]$ be the space of continuous functions $x\colon [0,\infty) \to \mathbb{R}$ such that the limit $\lim_{t\to\infty} x(t)$ exists and is finite. Similarly, let $C[-\infty,\infty]$ be the space of functions $y\colon (-\infty,\infty) \to \mathbb{R}$ such that both limits $\lim_{t\to\infty} y(t)$ and $\lim_{t\to-\infty} y(t)$ exist and are finite; we equip $C[-\infty,\infty]$ with the norm $\|y\| := \sup_{t\in\mathbb{R}} |y(t)|$. Also, let $\gamma \geq 0$ be a given number. For any $x \in C[0,\infty]$ we define its extension to a member y of $C[-\infty,\infty]$ as follows: we agree that $y(t) = x(t)$ for $t \geq 0$ and

$$y(-t) = x(t) - 2\gamma \int_0^t \mathrm{e}^{-\gamma(t-s)} x(s)\,\mathrm{d}s, \qquad t \geq 0.$$

1. Check to see that y is an element of $C[-\infty,\infty]$ whenever x belongs to $C[0,\infty]$.
2. Prove that $x \mapsto y$ is a bounded linear operator having norm not exceeding 3, provided $\gamma > 0$, and that its norm is one for $\gamma = 0$.

Exercise 12.10. Let $(\eta_i)_{n\geq 1}$ be a bounded sequence. Check to see that the formula

$$(\xi_i)_{i\geq 1} \mapsto \sum_{i=1}^{\infty} \xi_i \eta_i$$

defines a bounded linear functional on ℓ^1, and find its norm. (The series converges absolutely being dominated by $\sum_{i=1}^{\infty} M|\xi_i|$, where $M := \sup_{i\geq 1} |\eta_i| < \infty$.) Since a bounded sequence need not belong to ℓ^1, this example shows that an analogue of the theorem of Section 12.4 is not true in Banach spaces. That is, in general, linear functionals on a Banach space cannot be identified with elements of this space.

Exercise 12.11. ▲ Are all functionals $f\colon \ell^1 \to \mathbb{R}$ of the form given in the previous exercise? **Hint:** Try $\eta_i := f(e_i), i \geq 1$ for e_i defined in (10.1).

Exercise 12.12. ▲ Let $\left(a_{i,j}\right)_{i,j\in\mathbb{N}}$ be a matrix such that $\sup_{j\in\mathbb{N}} \sum_{i=1}^{\infty} |a_{j,i}| < \infty$. We know that then $A\colon \ell^1 \to \ell^1$ defined in (12.5) is a bounded linear operator. Given $(\eta_i)_{n\geq 1} \in \ell^\infty$, we may thus think of the map

$$\ell^1 \ni (\xi_i)_{i\geq 1} \mapsto \sum_{i=1}^{\infty} \eta_i \sum_{j=1}^{\infty} \xi_j a_{j,i} \in \mathbb{R}$$

because the series on the right converges absolutely. Since the answer to the question posed in the previous exercise is in the affirmative, there is a $(\varpi_i)_{n\geq 1} \in \ell^\infty$ such that the right-hand side above equals $\sum_{i=1}^{\infty} \varpi_i \xi_i$. The map $\ell^\infty \ni (\eta_i)_{n\geq 1} \mapsto (\varpi_i)_{n\geq 1} \in \ell^\infty$, denoted A^*, is an example of a *dual operator*. Express $(\varpi_i)_{n\geq 1}$ in terms of $\left(a_{i,j}\right)_{i,j\in\mathbb{N}}$ and $(\eta_i)_{n\geq 1}$, and check to see that $\|A^*\| = \|A\|$.

☞ CHAPTER SUMMARY

Bounded linear operators form a natural class of maps between normed linear spaces: they are, by definition, linear and continuous, so that, in other words, they preserve, at least to some extent, the linear and topological structures of the normed spaces involved. As it turns out, bounded linear operators from one normed linear space to another form a normed linear space themselves. Moreover, if the range space is complete, the space of operators is complete also, and is thus a Banach space. We note that we have already encountered examples of bounded linear operators on the previous pages of this book and discuss a score of new ones. Nor do we refrain from calculating norms of some of them.

13
Working in $\mathcal{L}(\mathbb{X})$

As we have seen in Chapter 12, the space of bounded linear operators from a Banach space $\mathbb{X}$ to a Banach space $\mathbb{Y}$ is itself a Banach space. This statement, besides its appealing beauty and symmetry, has profound consequences, some of which we will discuss in this chapter.

Before proceeding, we note that the case of $\mathbb{X} = \mathbb{Y}$ is of particular interest, since $\mathcal{L}(\mathbb{X})$ is not only a Banach space but also a Banach algebra with composition of operators as multiplication – in other words, the mathematical structure in $\mathcal{L}(\mathbb{X})$ is considerably richer than in the general space $\mathcal{L}(\mathbb{X}, \mathbb{Y})$. Indeed, by (12.10), for any A and B from $\mathcal{L}(\mathbb{X})$,

$$\|AB\| \leq \|A\| \, \|B\|.$$

The identity operator I, which, to recall, is defined by $Ix = x, x \in \mathbb{X}$, is a neutral element for the operation of composition:

$$A = IA = AI \qquad \text{for all } A \in \mathcal{L}(\mathbb{X}).$$

However, notably, $\mathcal{L}(\mathbb{X})$ is in general not commutative. For instance, the operators A and B of Example 12.7, that is, the operators of shift to the left and right in $\mathcal{L}(c_0)$, respectively, do not commute. For, it matters whether I shift a sequence to the left first and then to the right, or first to the right and then to the left. The second operation will leave the shifted vector unchanged, whereas the first will cause it to lose its first coordinate (this coordinate will be replaced by 0).

This chapter is devoted to a couple of applications of Theorem 12.14 in the case where $\mathbb{X} = \mathbb{Y}$ is a Banach space.

13.1 Existence of Inverses

It is one of the elementary and yet so useful consequences of the completeness of the algebra $\mathcal{L}(\mathbb{X})$ that all operators that are sufficiently close to the identity operator I are invertible. To be more specific, the condition

$$\|A - I\| < 1 \tag{13.1}$$

guarantees that there is an operator $B \in \mathcal{L}(\mathbb{X})$, typically denoted A^{-1} and termed the inverse, such that

$$BA = AB = I.$$

For the proof, let $q := \|A - I\| < 1$ and let's think of the operators

$$B_n := I + (I - A) + (I - A)^2 + \cdots + (I - A)^n,$$

where $(I - A)^i$ is the ith power of $I - A$, that is, the operator defined inductively as follows: $(I - A)^0 = I$ and $(I - A)^{i+1} = (I - A)^i(I - A)$. By the triangle inequality and (12.10),

$$\begin{aligned}\|B_n - B_m\| &= \left\|\sum_{k=m+1}^{n} (I - A)^k\right\| \le \sum_{k=m+1}^{n} \|(I - A)^k\| \le \sum_{k=m+1}^{n} q^k \\ &= q^{m+1}\frac{1 - q^{n-m}}{1 - q} \le \frac{q^{m+1}}{1 - q}\end{aligned}$$

for all $n > m$. Since $q < 1$, it follows that $(B_n)_{n\ge1}$ is a Cauchy sequence in $\mathcal{L}(\mathbb{X})$, and thus, since $\mathcal{L}(\mathbb{X})$ is complete, there is a limit, say B, of $(B_n)_{n\ge1}$. Next, a direct calculation shows that

$$(I - A)B_n = B_n(I - A) = B_{n+1} - I.$$

Therefore, letting $n \to \infty$ yields $B - AB = B - BA = B - I$ (see the first example of Section 12.3). This, however, proves our claim.

The following formula, a by-product of our analysis,

$$A^{-1} = \sum_{k=0}^{\infty}(I - A)^k, \tag{13.2}$$

is a counterpart of the well-known expansion into power series known in real analysis:

$$\frac{1}{x} = \frac{1}{1 - (1 - x)} = \sum_{k=0}^{\infty}(1 - x)^k;$$

the latter series, similarly to the operator power series, converges provided that the number x is close to 1: that is, provided that $|1-x|<1$. The series on the right-hand side of (13.2) is known as the Neumann series.[1]

13.2 Solving Equations

Before we explain how the result obtained above can be used to solve equations in Banach spaces, let's review some terminology. Let $\mathbb{X}$ be a Banach space, and let A be an operator belonging to $\mathcal{L}(\mathbb{X})$. We say that A has a right inverse if there is a $B \in \mathcal{L}(\mathbb{X})$ such that

$$ABy = y \qquad \text{for all } y \in \mathbb{X}.$$

The existence of a right inverse proves that A is *surjective*, that is, 'onto': for any $y \in \mathbb{X}$ there is an x (to wit, $x := By$), such that $Ax = y$. Hence, if a right inverse exists, equation

$$Ax = y, \tag{13.3}$$

in which y is given and x is searched for, has at least one solution.

For instance, the operator A of Example 12.7 has a right inverse equal to B of the same example – a sequence shifted first to the right and then to the left is in fact unchanged. In this context, it is worth stressing that there can be many right inverses for a single operator A. For, in the example at hand, both B_1 and B_2 defined above are right inverses for A:

$$B_1x = (\xi_1, \xi_1, \xi_2, \ldots), \qquad B_2x = \left(\sum_{i=1}^{\infty} \frac{1}{2^i}\xi_i, \xi_1, \xi_2, \ldots\right), \tag{13.4}$$

and the reader will be able to come up with scores of other examples of right inverses for A. This issue is related to the fact that (13.3) can have many solutions. To repeat, the existence of a right inverse speaks of the existence but not of the uniqueness of solutions.

It is the left inverse, defined below, that is related to the uniqueness of solutions. By definition, a left inverse for an $A \in \mathcal{L}(\mathbb{X})$ is an operator $B \in \mathcal{L}(\mathbb{X})$ such that

$$BAy = y \qquad \text{for all } y \in \mathbb{X}.$$

[1] Carl Gottfried Neumann, a German mathematician, should not be confused with John von Neumann, Hungarian-American mathematician, physicist, computer scientist, engineer and polymath (and complete genius).

It can be checked (see Exercise 13.5) that a left inverse is uniquely determined, that is, that there is at most one left inverse for an operator. Hence, it seems appropriate to speak of *the* left inverse.

As already stated, the existence of the left inverse guarantees the uniqueness of solutions to (13.3): if a solution exists, it is unique. For, if there are two solutions, say x_1 and x_2, we have

$$x_1 - x_2 = BAx_1 - BAx_2 = By - By = 0,$$

showing that $x_1 = x_2$. In other words, an operator that has a left inverse is *injective* (assigns different values to different arguments): condition $Ax = 0$ implies $x = BAx = B0 = 0$, showing that the kernel of A is trivial. It follows that $x_1 \neq x_2$ implies $Ax_1 \neq Ax_2$.

The operator A of Example 12.7 is the left inverse for the operator B defined there. Hence, as one can also easily see by oneself, B is injective. The operator B, however, is not surjective: for instance, the set of values of B does not contain the vector $(1, 0, 0, \dots)$. This example illustrates the fact that the existence of the left inverse does not imply the existence of the right inverse (and vice versa, as we have seen before).

We have, however, the following result: if the left and right inverses for an operator A exist, they coincide. To see this, suppose that B is the left inverse for A, and C is a right inverse for A. Then, for any $x \in \mathbb{X}$,

$$Bx = B(ACx) = (BA)Cx = Cx,$$

proving the claim.

The discussion presented above sheds more light on condition (13.1). Since this condition guarantees the existence of the left and right inverses of A, it guarantees also the existence and uniqueness of solutions to (13.3) for any $y \in \mathbb{X}$.

> If (13.1) holds, equation (13.3) has precisely one solution x for any y.

We complete this section with two examples.

13.1 Example (Cf. Exercise 3.14) We will show that, regardless of the choice of $y \in C[0, 1]$ and $\alpha \in \mathbb{R}$, there is a unique $x \in C[0, 1]$ such that

$$x(t) - \alpha \int_0^t x(s)\,\mathrm{d}s = y(t), \qquad t \in [0, 1]. \tag{13.5}$$

To this end, we introduce the operator $A \in \mathcal{L}(\mathbb{X})$ by $Ax(t) = x(t) - \alpha \int_0^t x(s)\,\mathrm{d}s, t \in [0,1]$. Then (13.5) takes the form

$$Ax = y,$$

and our task reduces to checking that condition (13.1) is satisfied. Unfortunately, it turns out that in the standard maximum norm (13.1) does not hold, unless $|\alpha|$ is small. We are not discouraged, however, because (as in the second, better proof of the local theorem of Picard) we have an equivalent, Bielecki-type norm at our disposal:[2]

$$\|x\| := \max_{t\in[0,1]} \mathrm{e}^{-\lambda t}|x(t)|,$$

where $\lambda > 0$ is to be determined later. Since $(A - I)x(t) = -\alpha \int_0^t x(s)\,\mathrm{d}s$, we can thus write

$$\begin{aligned}
\|(A-I)x\| &= \sup_{t\in[0,1]} |\alpha|\mathrm{e}^{-\lambda t}\left|\int_0^t x(s)\,\mathrm{d}s\right| \le \sup_{t\in[0,1]} |\alpha|\mathrm{e}^{-\lambda t}\int_0^t |x(s)|\,\mathrm{d}s \\
&= \sup_{t\in[0,1]} |\alpha| \int_0^t \mathrm{e}^{-\lambda(t-s)}\mathrm{e}^{-\lambda s}|x(s)|\,\mathrm{d}s \le \sup_{t\in[0,1]} |\alpha| \int_0^t \mathrm{e}^{-\lambda(t-s)}\|x\|\,\mathrm{d}s \\
&= \sup_{t\in[0,1]} |\alpha| \int_0^t \mathrm{e}^{-\lambda s}\,\mathrm{d}s\,\|x\| < |\alpha| \int_0^\infty \mathrm{e}^{-\lambda s}\,\mathrm{d}s\,\|x\| \\
&= \frac{|\alpha|}{\lambda}\|x\|.
\end{aligned}$$

Hence, condition (13.1) holds in the Bielecki-type norm as long as $\lambda > |\alpha|$. This proves our claim, because the Bielecki-type norm, being equivalent to the standard maximum norm, makes the space $C[0,1]$ complete. □

A short calculation shows that

$$(I-A)^k y(t) = \alpha^k \int_0^t \frac{(t-s)^{k-1}}{(k-1)!} y(s)\,\mathrm{d}s, \qquad t \in [0,1], k \ge 1.$$

Therefore, (13.2) leads to the conjecture that the solution to (13.5) is given by

$$x(t) = y(t) + \alpha \int_0^t \mathrm{e}^{\alpha(t-s)} y(s)\,\mathrm{d}s, \qquad t \in [0,1], \tag{13.6}$$

and one can check that this x indeed solves (13.5). In general, though, (13.2) rarely gives an explicit form of solutions to (13.3); all condition (13.1) does, therefore, is primarily to guarantee the existence and uniqueness of these solutions. As our second example illustrates, sometimes from (13.2) one can at least deduce useful information about the solutions.

[2] To the reader who contemplates the fact that in one norm the theorem of the previous section does not work, and in another norm it does, I hasten to add that (13.1) is a sufficient, but not necessary, condition for the existence and uniqueness of solutions.

13.2 Example Suppose $\gamma_{i,j}, i, j = 1, \dots, k$ are non-negative numbers such that

$$\sum_{j=1}^{k} \gamma_{i,j} < 1 \qquad \text{for all } i = 1, \dots, k.$$

Then, for any $\eta_1, \dots, \eta_k \in \mathbb{R}$, the system

$$\xi_j - \sum_{i=1}^{k} \gamma_{i,j}\xi_i = \eta_j \qquad \text{for all } j = 1, \dots, k \tag{13.7}$$

has precisely one solution, $x = (\xi_1, \dots, \xi_k) \in \mathbb{R}^k$. Moreover, $x \geq 0$ provided $y = (\eta_1, \dots, \eta_k) \geq 0$.

Again, as in Example 2.4, this seems to be a problem in linear algebra, but I am afraid using Cramer's rule is next to impossible here. However, our method will work smoothly. For $\mathbb{X}$ we take $\mathbb{R}^k$, equipped with the norm $\|x\| = \sum_{i=1}^{k} |\xi_i|$, and let $B \in \mathcal{L}(\mathbb{R}^k)$ be the operator given by

$$B(\xi_i)_{i=1,\dots,k} = \left(\sum_{i=1}^{k} \gamma_{i,j}\xi_i \right)_{j=1,\dots,k}.$$

(This is multiplying the matrix $B = (\gamma_{i,j})_{i,j=1,\dots k}$ by a row-vector from the left, as in (12.1).) Then

$$\begin{aligned} \|B\| &\leq \sup_{\|x\|=1} \sum_{j=1}^{k} \sum_{i=1}^{k} \gamma_{i,j} |\xi_i| = \sup_{\|x\|=1} \sum_{i=1}^{k} |\xi_i| \sum_{j=1}^{k} \gamma_{i,j} \\ &\leq \gamma := \max_{i=1,\dots,k} \sum_{j=1}^{k} \gamma_{i,j}. \end{aligned}$$

Since, by assumption, $\gamma < 1$, for $A := I - B$, we have $\|I - A\| = \|B\| \leq \gamma < 1$. It follows that there is precisely one solution to $Ax = y$, that is, precisely one solution to (13.7). Moreover, by (13.2), $x = \sum_{n=0}^{\infty} B^n y$. Therefore, since the operators B^n map non-negative vectors of $\mathbb{R}^k$ into non-negative vectors of $\mathbb{R}^k$, we are done.

13.3 Exponential Function for a Bounded Operator

Completeness of the algebra $\mathcal{L}(\mathbb{X})$ allows us also to prove the existence of an exponential function of a bounded operator, as we will now explain. Notably, the exponential function is counted among the most important functions of contemporary analysis, if not of the whole of mathematics (see the Prologue to [35]). We remember that, for a real (or even complex) a the function $\mathbb{R} \ni t \mapsto \mathrm{e}^{ta}$ is defined as

$$\mathrm{e}^{ta} = \sum_{k=0}^{\infty} \frac{t^k a^k}{k!}.$$

Hence, given an operator $A \in \mathcal{L}(\mathbb{X})$ and $t \in \mathbb{R}$, we consider sums:

$$S_n(t) = \sum_{k=0}^{n} \frac{t^k A^k}{k!}. \tag{13.8}$$

Of course, our hope is that the sequence $(S_n(t))_{n\geq 1}$ converges in $\mathcal{L}(\mathbb{X})$, so that we could define e^{tA} as its limit.

Without the information that $\mathcal{L}(\mathbb{X})$ is complete, we would have come to a dead end here. For, how can we prove that a sequence converges? If we want to show that something is its limit, we need to check that the distances of elements of this sequence from this something converge to 0, and this cannot be done if this something is not given, if its existence has not been established.

We are thus in dire straits, and it is completeness that comes to our aid. To wit, completeness allows us to ascertain the existence of the limit by merely examining distances between elements of the sequence itself. To this end, we think of $n > m$ and write

$$\begin{aligned} \|S_n - S_m\| &= \left\| \sum_{k=m+1}^{n} \frac{t^k A^k}{k!} \right\| \leq \sum_{k=m+1}^{n} \left\| \frac{t^k A^k}{k!} \right\| \leq \sum_{k=m+1}^{n} \frac{|t|^k \|A^k\|}{k!} \\ &\leq \sum_{k=m+1}^{n} \frac{|t|^k \|A\|^k}{k!}, \end{aligned}$$

where the triangle inequality and (12.10) were used. The sum on the right-hand side converges to zero, as $m, n \to \infty$, because it is the difference between the nth and the mth sum of the e^b defining series for the real number $b = |t|\,\|A\|$. And this is it: completeness renders the existence of the limit of $(S_n(t))_{n\geq 1}$ and allows us to define:

$$\mathrm{e}^{tA} = \sum_{k=0}^{\infty} \frac{t^k A^k}{k!}.$$

Without completeness we could not be sure that this infinite sum exists.

13.3 Example Let $\mathbb{X} = \mathbb{R}^2$ and let A be given the matrix $A = \begin{pmatrix} -a & a \\ b & -b \end{pmatrix}$ in which $a \geq 0$ and $b \geq 0$ satisfying $a + b = 1$ are given. We will show that

$$e^{tA} = \begin{pmatrix} b + ae^{-t} & a - ae^{-t} \\ b - be^{-t} & be^{-t} + a \end{pmatrix}, \qquad t \in \mathbb{R}. \tag{13.9}$$

To this end, we calculate the characteristic polynomial of the matrix A to be $\lambda(\lambda + a + b) = \lambda(\lambda + 1)$. Hence, there are two characteristic values: 0 and -1. Next, we find the corresponding characteristic vectors and conclude that A has the following diagonal form:

$$A = \begin{pmatrix} 1 & -a \\ 1 & b \end{pmatrix} \begin{pmatrix} 0 & 0 \\ 0 & -1 \end{pmatrix} \begin{pmatrix} b & a \\ -1 & 1 \end{pmatrix}.$$

This shows that

$$A^i = \begin{pmatrix} 1 & -a \\ 1 & b \end{pmatrix} \begin{pmatrix} 0 & 0 \\ 0 & (-1)^i \end{pmatrix} \begin{pmatrix} b & a \\ -1 & 1 \end{pmatrix}, \qquad i \geq 1,$$

and

$$\sum_{i=0}^{n} \frac{t^i}{i!} A^i = \begin{pmatrix} 1 & -a \\ 1 & b \end{pmatrix} \begin{pmatrix} 1 & 0 \\ 0 & \sum_{i=0}^{n} \frac{t^i (-1)^i}{i!} \end{pmatrix} \begin{pmatrix} b & a \\ -1 & 1 \end{pmatrix}, \qquad n \geq 1.$$

This is but just one step away from

$$e^{tA} = \begin{pmatrix} 1 & -a \\ 1 & b \end{pmatrix} \begin{pmatrix} 1 & 0 \\ 0 & e^{-t} \end{pmatrix} \begin{pmatrix} b & a \\ -1 & 1 \end{pmatrix}.$$

By multiplying out these matrices we obtain (13.9). □

We will see more examples of exponential functions in what follows.

13.4 Differential Equations in Banach Spaces

Among the reasons that make the exponential function so special and important is the fact that this function describes solutions to differential equations. To wit, the function u defined by $u(t) := e^{ta}x, t \in \mathbb{R}$ is the unique solution to the differential equation

$$u'(t) = au(t), \qquad u(0) = x,$$

where a is a given real number, and $x \in \mathbb{R}$ is an initial condition. It turns out that the operator-valued exponential function has similar properties, and this allows us to solve difficult equations in a simple way.

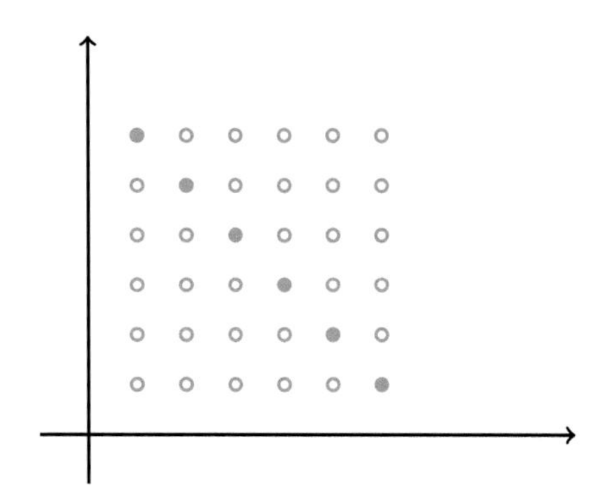

Figure 13.1 In (13.10) we sum over the entire 'square' and in (13.11) only over its lower part, that is, over the 'triangle' lying below the diagonal marked by filled circles.

We will make this statement clearer in a series of lemmas that culminate in Theorem 13.7. Throughout this section $\mathbb{X}$ is a Banach space, and A is a bounded linear operator in $\mathbb{X}$, that is, $A \in \mathcal{L}(\mathbb{X})$.

13.4 Lemma *We have*

$$\mathrm{e}^{tA}\mathrm{e}^{sA} = \mathrm{e}^{(s+t)A} \qquad \textit{for all } s,t \in \mathbb{R}.$$

Proof Let's fix a natural n, and consider the product of two partial sums:

$$S_n(t)S_n(s) = \sum_{k=0}^{n} \frac{t^k A^k}{k!} \sum_{k=0}^{n} \frac{s^k A^k}{k!} = \sum_{0 \le k,l \le n} \frac{t^k s^l A^{k+l}}{k!\, l!}. \tag{13.10}$$

As $n \to \infty$, this product converges to $\mathrm{e}^{tA}\mathrm{e}^{sA}$ (see Section 12.3). Therefore, we consider a similar sum (see Figure 13.1):

$$R_n := \sum_{\substack{0 \le k,l \le n \\ 0 \le k+l \le n}} \frac{t^k s^l A^{k+l}}{k!\, l!}. \tag{13.11}$$

Since for positive numbers condition $k + l \le n$ implies $k,l \le n$,

$$R_n = \sum_{k,l \ge 0, k+l \le n} \frac{t^k s^l A^{k+l}}{k!\, l!}.$$

This in turn is the same as

$$\sum_{m=0}^{n} \sum_{k=0}^{m} \frac{t^k s^{m-k} A^m}{k!\,(m-k)!};$$

to see this, think of m as being equal to $k + l$; in effect, we sum over the diagonals that are parallel to the one marked by filled circles in Figure 13.1. By Newton's binomial formula (see e.g., [20]):

$$\sum_{k=0}^{m} \frac{t^k s^{m-k}}{k!\,(m-k)!} = \frac{(t+s)^m}{m!};$$

we thus obtain

$$R_n = \sum_{m=0}^{n} \frac{(t+s)^m A^m}{m!}.$$

This means that R_n coincides with $S_n(t+s)$, proving that $\lim_{n\to\infty} R_n = \mathrm{e}^{(t+s)A}$.

We are therefore left with proving that the difference of $S_n(t)S_n(s)$ and R_n tends to 0, as $n \to \infty$. By the triangle inequality, the norm of this difference does not exceed

$$\sum_{\substack{0 \le k,l \le n \\ k+l \ge n+1}} \left\| \frac{t^k s^l A^{k+l}}{k!\,l!} \right\| \le \sum_{k+l \ge n+1} \left\| \frac{t^k s^l A^{k+l}}{k!\,l!} \right\| \le \sum_{k+l=n+1}^{\infty} \frac{|t|^k |s|^l \|A\|^{k+l}}{k!\,l!};$$

the second sum here is larger or equal to the first because it has more (non-negative) terms, and the third is larger or equal to the second by (12.10). Summing over diagonals as above, we obtain that the third sum equals

$$\sum_{m=n+1}^{\infty} \sum_{k=0}^{m} \frac{|t|^k |s|^{m-k} \|A\|^m}{k!\,(m-k)!} = \sum_{m=n+1}^{\infty} \frac{(|t|+|s|)^m \|A\|^m}{m!}.$$

The last expression tends to 0, as $n \to \infty$, being the difference between the full and partial sums of the numerical series defining $\mathrm{e}^{(|t|+|s|)\|A\|}$. This completes the proof. □

13.5 Lemma *Functions $t \mapsto \mathrm{e}^{tA}$ and $t \mapsto A\mathrm{e}^{tA} = \mathrm{e}^{tA}A$ are continuous.*

Proof Let's start from $t \mapsto \mathrm{e}^{tA}$.

1. For each n, $S_n(t)$ of (13.8) depends continuously on t, being a polynomial in t.

2. Let $T > 0$ be fixed. We have

$$\sup_{t\in[-T,T]} \|S_n(t) - \mathrm{e}^{tA}\| = \sup_{t\in[-T,T]} \left\| \sum_{k=n+1}^{\infty} \frac{t^k A^k}{k!} \right\|$$
$$\leq \sup_{t\in[-T,T]} \sum_{k=n+1}^{\infty} \frac{|t|^k \|A\|^k}{k!} \leq \sum_{k=n+1}^{\infty} \frac{T^k \|A\|^k}{k!}.$$

   Arguing as in the last part of the proof of Lemma 13.4, we check that the last expression converges to 0, as $n \to \infty$. This shows that in the interval $[-T,T]$, $S_n(t)$ converges to e^{tA} uniformly with respect to t.
3. By points 1. and 2. above, $t \mapsto \mathrm{e}^{tA}$ is continuous in $[-T,T]$, as a uniform limit of continuous functions. Since T is arbitrary, we are done.

Turning to the second part, we note first that

$$AS_n(t) = A\sum_{k=0}^{n} \frac{t^k A^k}{k!} = \sum_{k=0}^{n} \frac{t^k A^{k+1}}{k!} = \sum_{k=0}^{n} \frac{t^k A^k}{k!} A.$$

Letting $n \to \infty$, we obtain $A\mathrm{e}^{tA} = \mathrm{e}^{tA}A$. The rest is an immediate consequence of the continuity of $t \mapsto \mathrm{e}^{tA}$. Indeed, we have $\|A\mathrm{e}^{tA} - A\mathrm{e}^{sA}\| \leq \|A\|\|\mathrm{e}^{tA} - \mathrm{e}^{sA}\|$. □

13.6 Lemma *For each t,*

$$I + \int_0^t A\mathrm{e}^{sA}\,\mathrm{d}s = \mathrm{e}^{tA}.$$

As a result, $t \mapsto \mathrm{e}^{tA}$ is continuously differentiable with

$$\frac{\mathrm{d}}{\mathrm{d}t}\mathrm{e}^{tA} = A\mathrm{e}^{tA} = \mathrm{e}^{tA}A.$$

Proof By Exercise 6.3, we can write

$$I + \int_0^t AS_n(s)\,\mathrm{d}s = I + \int_0^t \sum_{k=0}^{n} \frac{s^k A^{k+1}}{k!}\,\mathrm{d}s = I + \sum_{k=0}^{n} \int_0^t \frac{s^k A^{k+1}}{k!}\,\mathrm{d}s$$
$$= I + \sum_{k=0}^{n} \frac{t^{k+1} A^{k+1}}{(k+1)!} = S_{n+1}(t).$$

Therefore, to obtain the first part of our lemma, it suffices to show that $\int_0^t AS_n(s)\,\mathrm{d}s$ converges to $\int_0^t A\mathrm{e}^{sA}\,\mathrm{d}s$, as $n \to \infty$. We argue as above: the norm of the difference between these two integrals is

$$\begin{aligned}\left\|\int_0^t A\,[S_n(s) - \mathrm{e}^{sA}]\,\mathrm{d}s\right\| &\le \|A\| \left|\int_0^t \|S_n(s) - \mathrm{e}^{sA}\|\,\mathrm{d}s\right| \\ &\le \|A\|\,|t| \sup_{s\in[0,|t|]} \sum_{k=n+1}^{\infty} \frac{|s|^k \|A\|^k}{k!} \\ &\le \|A\|\,|t| \sum_{k=n+1}^{\infty} \frac{|t|^k \|A\|^k}{k!};\end{aligned}$$

the first inequality here is a result of (6.8) and (12.10). (Note that t above might be negative; by convention, for $t < 0$, $\int_0^t := -\int_t^0$.) The last expression tends to 0 as $n \to \infty$, as the difference between the full and the partial sums of the numerical series defining $\mathrm{e}^{|t|\,\|A\|}$.

By the first part, for $t \in \mathbb{R}$ and $h \neq 0$,

$$\frac{\mathrm{e}^{(t+h)A} - \mathrm{e}^{tA}}{h} = \frac{1}{|h|} \int_t^{t+h} A\mathrm{e}^{sA}\,\mathrm{d}s.$$

Hence,

$$\frac{\mathrm{e}^{(t+h)A} - \mathrm{e}^{tA}}{h} - A\mathrm{e}^{tA} = \frac{1}{|h|} \int_t^{t+h} [A\mathrm{e}^{sA} - A\mathrm{e}^{tA}]\,\mathrm{d}s.$$

Since, by Lemma 13.5, $t \mapsto A\mathrm{e}^{sA}$ is continuous, having given $\epsilon > 0$, we can choose an h so that the norm of the integrand on the right-hand side does not exceed ϵ for s in the interval of integration (whether $[t, t+h]$ for $h > 0$ or $[t+h, t]$ for $h < 0$). Hence, for such an h, the norm of the left-hand side does exceed

$$\frac{1}{|h|} \times (\text{max. norm of the integrand}) \times (\text{interval's length}) \le \frac{1}{|h|}\epsilon |h| = \epsilon.$$

This completes the proof. □

We are finally able to exhibit the fundamental connection between exponential functions and solutions to initial value problems (i.e., Cauchy problems) for linear, autonomous differential equations in Banach spaces.

13.7 Theorem *Let x be a vector in a Banach space $\mathbb{X}$, and let $A \in \mathcal{L}(\mathbb{X})$. For any $T > 0$, the differential equation*

$$u'(t) = Au(t), \qquad \text{with initial condition } u(0) = x, \quad t \in [-T, T] \quad (13.12)$$

has precisely one solution given by $u(t) := \mathrm{e}^{tA}x, t \in [-T, T]$.

Proof By Lemma 13.6, the function defined above is a solution to our equation. What we need to show is that there are no other solutions. To this end, we assume that $[0, T] \ni t \mapsto u(t)$ is a solution to our equation, fix an

$s \in (0, T]$, and consider the function $t \mapsto v(t) = e^{(s-t)A}u(t)$, defined for $t \in [0, s]$. Since

$$\frac{d}{dt}e^{(s-t)A} = e^{sA}\frac{d}{dt}e^{-tA} = e^{sA}\frac{d}{dt}e^{t(-A)} = e^{sA}(-A)e^{t(-A)} = -Ae^{(s-t)A},$$

this function is differentiable, and its derivative is

$$\begin{aligned}\frac{d}{dt}\left[e^{(s-t)A}u(t)\right] &= -Ae^{(s-t)A}u(t) + e^{(s-t)A}u'(t)\\ &= -Ae^{(s-t)A}u(t) + e^{(s-t)A}Au(t).\end{aligned}$$

The relation $e^{(s-t)A}A = Ae^{(s-t)A}$ implies that the derivative is zero. Hence, v is constant (see Appendix). It follows that $u(s) = Iu(s) = v(s) = v(0) = e^{sA}u(0) = e^{sA}x$ ($u(0) = x$ because u solves our differential equation with prescribed boundary condition). Since $s \in (0, T]$ is arbitrary, we see that $u(s) = e^{sA}x$ for $s \in [0, T]$. The case of $s \in [-T, 0)$ is similar: it suffices to consider v on the interval $[s, 0]$, and therefore we omit the details. □

13.8 Example The matrix

$$A = \begin{pmatrix} -a & a \\ b & -b \end{pmatrix},$$

in which a and b are positive numbers, is an example of an intensity (or Kolmogorov) matrix known in the theory of continuous time Markov chains. This, by definition, simply means that entries in either of its rows add up to 0 and that the off-diagonal entries are non-negative. The related Markov chain obeys the following rules. It has two states, say 1 and 2; at 1 it stays for an exponential time, say, T, with parameter a (i.e., for all $t \geq 0$, $\Pr(T > t) = e^{-at}$) and then jumps to 2. At 2, on the other hand, it stays for an independent exponential time with parameter b before jumping back to 1 where the story begins anew, with new, independent exponential times.

By combining (13.9) with Exercise 13.12 we obtain that the exponent of A is

$$e^{tA} = \frac{1}{a+b}\begin{pmatrix} b + ae^{-(a+b)t} & a - ae^{-(a+b)t} \\ b - be^{-(a+b)t} & be^{-(a+b)t} + a \end{pmatrix}, t \in \mathbb{R}.$$

These matrices gather transition probabilities for the Markov chain described above. To wit, the upper left corner is the probability that the chain starting at 1 will be there also at time $t > 0$, and the upper right corner is the probability that this chain will be at this time at 2. The second row is interpreted similarly, that is, it describes the process that starts at 2. We note that for $t < 0$ this interpretation does not make sense.

Solutions to the differential equation

$$u'(t) = Au(t), \qquad u(0) = (\xi_1, \xi_2),$$

can also be interpreted in probabilistic terms. To this end, we need, similarly as in Example 12.4, to equip $\mathbb{R}^2$ with the norm $\|(\xi_1, \xi_2)\| = |\xi_1| + |\xi_2|$ and identify the matrix A with the operator given by

$$A(\xi_1, \xi_2) = (\xi_1, \xi_2)A$$

(matrix A is multiplied here by a row-vector from the left). Then, as long as $\xi_1, \xi_2 \geq 0$ and $\xi_1 + \xi_2 = 1$, the first coordinate of

$$u(t) := (\xi_1, \xi_2)\mathrm{e}^{tA}, \tag{13.13}$$

that is, the quantity

$$\xi_1 \left(\frac{b}{a+b} + \frac{a}{a+b}\mathrm{e}^{-(a+b)t} \right) + \xi_2 \left(\frac{b}{a+b} - \frac{b}{a+b}\mathrm{e}^{-(a+b)t} \right),$$

is the probability that the Markov chain described above that starts at 1 with probability ξ_1 and starts at 2 with probability ξ_2 will be at 1 at time $t \geq 0$. Again, for $t < 0$, this interpretation is impossible.

13.9 Example Let's solve the ordinary differential equation

$$u''(t) = -\alpha^2 u(t), \qquad t \geq 0, \tag{13.14}$$

with initial conditions $u(0) = \xi_1, u'(0) = \xi_2$. We are looking for a real-valued u here, and α is a given real number; to avoid trivialities we assume $\alpha \neq 0$. This is an equation of second order and because of this apparently does not fit in with Theorem 13.7. But we can change it into an equivalent equation of first order in $\mathbb{R}^2$. Namely, by introducing $v = u'$, we transform (13.14) into

$$\begin{cases} u'(t) = v(t), \\ v'(t) = u''(t) = -\alpha u(t) \end{cases}$$

or, equivalently,

$$\begin{pmatrix} u(t) \\ v(t) \end{pmatrix}' = A \begin{pmatrix} u(t) \\ v(t) \end{pmatrix} \tag{13.15}$$

for $A := \begin{pmatrix} 0 & 1 \\ -\alpha^2 & 0 \end{pmatrix}$ (here, A is multiplied by a column vector from the right). By Theorem 13.7 the unique solution to this equation is

$$\begin{pmatrix} u(t) \\ v(t) \end{pmatrix} = \mathrm{e}^{tA} \begin{pmatrix} \xi_1 \\ \xi_2 \end{pmatrix}, \qquad t \geq 0.$$

To find e^{tA} explicitly, we note that

$$A^2 = \begin{pmatrix} 0 & 1 \\ -\alpha^2 & 0 \end{pmatrix} \begin{pmatrix} 0 & 1 \\ -\alpha^2 & 0 \end{pmatrix} = -\alpha^2 \begin{pmatrix} 1 & 0 \\ 0 & 1 \end{pmatrix} = -\alpha^2 I,$$

which, in turn, implies by induction that

$$A^{2n} = (-\alpha^2)^n I \qquad \text{and} \qquad A^{2n+1} = (-\alpha^2)^n A, \qquad n \ge 1.$$

It follows that

$$\begin{aligned} e^{tA} &= \sum_{n=0}^{\infty} \frac{t^{2n}}{(2n)!} A^{2n} + \sum_{n=0}^{\infty} \frac{t^{2n+1}}{(2n+1)!} A^{2n+1} \\ &= \sum_{n=0}^{\infty} \frac{t^{2n}(-\alpha^2)^n}{(2n)!} I + \sum_{n=0}^{\infty} \frac{t^{2n+1}(-\alpha^2)^n}{(2n+1)!} A = \cos \alpha t I + \frac{\sin \alpha t}{\alpha} A. \end{aligned}$$

In sum,

$$\begin{pmatrix} u(t) \\ v(t) \end{pmatrix} = \begin{pmatrix} \cos \alpha t & \frac{\sin \alpha t}{\alpha} \\ -\alpha \sin \alpha t & \cos \alpha t \end{pmatrix} \begin{pmatrix} \xi_1 \\ \xi_2 \end{pmatrix}.$$

Hence, the solution to (13.14) is

$$u(t) = \xi_1 \cos \alpha t + \xi_2 \frac{\sin \alpha t}{\alpha}$$

(we do not need the formula for v, because v, as the derivative of u, can be recovered from u).

13.10 Example Let $\mathbb{X}$ be the space $C[0,\infty]$ of real continuous functions on $[0,\infty)$ with finite limits at infinity, and let the operators A and B in $\mathcal{L}(\mathbb{X})$ be given by

$$A = B - \lambda I, \quad Bx(s) = \lambda x(s+1), \qquad s \in [0,\infty];$$

$\lambda > 0$ is a given real number here. By the second part of Exercise 13.12,

$$e^{tA} = e^{-\lambda t} e^{tB}.$$

Since $B^n x(s) = \lambda^n x(s+n)$, we have also

$$e^{tB}x(s) = \sum_{n=0}^{\infty} \frac{\lambda^n t^n}{n!} x(s+n) \qquad \text{and} \qquad e^{tA}x(s) = \sum_{n=0}^{\infty} e^{-\lambda t} \frac{\lambda^n t^n}{n!} x(s+n) \tag{13.16}$$

for $s \ge 0, t \ge 0$. To recall (see e.g., [20]), $e^{-\lambda t} \frac{\lambda^n t^n}{n!}$ is the probability that a Poisson process $\{N(t), t \ge 0\}$ with intensity λ takes the value n at the time $t \ge 0$. Hence, for $t \ge 0$, the solution to the differential equation

$$u'(t) = Au(t), \qquad u(0) = x,$$

in $C[0,\infty]$ can be written in the following compact form:

$$[u(t)](s) = \mathrm{e}^{tA}x(s) = \sum_{n=0}^{\infty} \Pr(N(t)=n)x(s+n) = E\,x(s+N(t)),\ s \geq 0, t \geq 0$$

(E denotes expected value here). Again, as in Example 13.8, the last formula makes sense for $t \geq 0$, whereas (13.16) holds also for $t < 0$.

13.5 Exercises

In the following exercises, unless otherwise stated, $\mathbb{X}$ is a Banach space.

Exercise 13.1. Let $P \in \mathcal{L}(\mathbb{X})$ be *idempotent*, that is, $P^2 = P$. For an $a \neq -1$, find $(I + aP)^{-1}$. Check to see that $\lim_{a\to\infty}(I + aP)^{-1} = I - P$. Conclude also that small distance from I is far from being a necessary condition for invertibility. **Hint:** For the inverse search among linear combinations of I and P.

Exercise 13.2. Check that for any $\epsilon > 0$ and integer n, the $n \times n$ matrix $A(\epsilon) = (a_{i,j}(\epsilon))_{i,j=1,\dots,n}$ where $a_{i,i}(\epsilon) = (1 + \epsilon)$ and $a_{i,j}(\epsilon) = 1$ for $j \neq i, i = 1, \dots, n$ is invertible and find the limit $\lim_{\epsilon\to 0} \epsilon[A(\epsilon)]^{-1}$. **Hint:** Use the previous exercise.

Exercise 13.3. Let $P \in \mathcal{L}(\mathbb{X})$ be such that $P^3 = P$. For an $a \neq \pm 1$, find $(I + aP)^{-1}$. Check to see that $\lim_{a\to\infty}(I + aP)^{-1} = I - P^2$.

Exercise 13.4. Let $A \in \mathcal{L}(\mathbb{X})$. For $n > \|A\|$, $I - \frac{1}{n}A$ is invertible. Check that $\lim_{n\to\infty}(I - \frac{1}{n}A)^{-1} = I$. **Hint:** By (13.2), $\|\frac{1}{n}(I - \frac{1}{n}A)^{-1}\| \leq \frac{1}{n-\|A\|}$; write $(I - \frac{1}{n}A)^{-1} - I = \frac{1}{n}(I - \frac{1}{n}A)^{-1}A$. (Compare Lemma 16.10.)

Exercise 13.5. Check that for any $A \in \mathcal{L}(\mathbb{X})$ there is at most one left inverse operator.

Exercise 13.6. Let $\mathbb{X}$ be the space $C[0,1]$ of real continuous functions on $[0,1]$, and let $A \in \mathcal{L}(\mathbb{X})$ be given by

$$Ax(t) = \frac{x(t)}{1+t}, \qquad t \in [0,1], x \in \mathbb{X}.$$

Check to see that A is invertible with inverse

$$A^{-1}x(t) = (1+t)x(t), \qquad t \in [0,1], x \in \mathbb{X}.$$

Also $\|A\| = 1$ and $\|A^{-1}\| = 2$. Conclude that the inequality in (12.10) can be strict.

Exercise 13.7. Suppose $\mathbb{X}$ is a Banach algebra and there is a neutral element for multiplication in $\mathbb{X}$, termed unity, and denoted $\mathfrak{u}$: this means that for all $a \in \mathbb{X}$, $a\mathfrak{u} = \mathfrak{u}a = a$. Show that for $a \in \mathbb{X}$ condition $\|a - \mathfrak{u}\| < 1$ implies the existence of an inverse of a, that is, the existence of an element of $\mathbb{X}$, denoted a^{-1}, such that $a^{-1}a = aa^{-1} = \mathfrak{u}$. Furthermore, $\|a^{-1} - \mathfrak{u}\| \le \frac{\|a-\mathfrak{u}\|}{1-\|a-\mathfrak{u}\|}$. More generally, if a has the inverse a^{-1}, and b lies so close to a that $\|a-b\| < \frac{1}{\|a^{-1}\|}$, then b also has an inverse, and

$$\|a^{-1} - b^{-1}\| \le \frac{\|a^{-1}\|^2\|a - b\|}{1 - \|a^{-1}\|\|a - b\|}.$$

Exercise 13.8. Prove that the operators B_1 and B_2 defined by (13.4) are bounded and find their norms. Find yet another right inverse for the operator A from the example in question.

Exercise 13.9. Let g be a given function in $C[0, 1]$ such that $g(1) = 0$. For any $c \in \mathbb{R}$, let A_c be given by $A_c f = f - cf(1)g$, $f \in C[0, 1]$. Check to see that A_{-c} is the left and right inverse for A_c. **Hint:** $A_c f(1) = f(1)$.

Exercise 13.10. Suppose that, as in Example 13.2, $\gamma_{i,j}, i, j = 1, \dots, k$ are non-negative numbers such that

$$\sum_{j=1}^{k} \gamma_{i,j} < 1 \qquad \text{for all } i = 1, \dots, k.$$

Check that for any $\eta_1, \dots, \eta_k \in \mathbb{R}$, the system

$$\xi_i - \sum_{j=1}^{k} \gamma_{i,j}\xi_j = \eta_i \qquad \text{for all } i = 1, \dots, k$$

has precisely one solution $x = (\xi_1, \dots, \xi_k) \in \mathbb{R}^k$. Moreover, $x \ge 0$ provided $y = (\eta_1, \dots, \eta_k) \ge 0$. **Hint:** Argue as in Example 13.2, but equip $\mathbb{R}^k$ with the maximum norm.

Exercise 13.11. Check that x given by (13.6) solves (13.5).

Exercise 13.12. Prove that for any real number a and any bounded linear operator A, exponential functions for A, $A - aI$ and aA are related as follows:

$$\mathrm{e}^{t(aA)} = \mathrm{e}^{atA} \qquad \text{and} \qquad \mathrm{e}^{t(A-aI)} = \mathrm{e}^{-at}\mathrm{e}^{tA}, \qquad t \in \mathbb{R}.$$

Hint: To prove the second formula use Theorem 13.7.

Exercise 13.13. Here is the idea for an alternative proof of (13.9). Since the characteristic polynomial of A is $p(\lambda) = \lambda(\lambda + 1)$, the Cayley–Hamilton theorem tells us that $A^2 = -A$. This allows us to conclude that $A^n = (-1)^{n+1}A$ for $n \geq 1$ and so $e^{tA} = I + A - e^{-t}A$, from which (13.9) follows. Complete the details of this reasoning.

Exercise 13.14. Let $n \geq 2$ be an integer, and let $Q = (q_{i,j})_{i,j=1,\dots,n}$ be an $n \times n$ intensity matrix; this means that all its off-diagonal entries are non-negative, and the sum of the entries in each row is zero. Prove that $e^{tQ}, t \geq 0$ is composed of stochastic matrices, that is, that all their entries are non-negative and the sum of the entries in each row is 1. **Hint:** Write $Q = Q + qI - qI$ where I is the identity matrix and $q := \max_{i=1,\dots,n}(-q_{i,i})$, and use Exercise 13.12.

Exercise 13.15. Let A be a bounded linear operator. Prove that

(a) If $A^2 = 0$, then $e^{tA} = I + tA, t \in \mathbb{R}$.
(b) If $A^2 = A$, then $e^{tA} = I + (e^t - 1)A, t \in \mathbb{R}$.

Exercise 13.16. Let $\chi \in C[0,1]$ be such that $\chi(1) = 0$. Use the previous exercise to find e^{tA} for the bounded linear operator in $C[0,1]$ given by $Ax = x(1)\chi$.

Exercise 13.17. Prove the converse to (a) and (b) in Exercise 13.15. **Hint for (b):** If $t \mapsto I + (e^t - 1)A$ is an exponent, then $[I + (e^t - 1)A][I + (e^t - 1)A] = I + (e^{2t} - 1)A$.

Exercise 13.18. ▲ Let A and B be two bounded linear operators in the same Banach space. Prove that if $B^2 = B$ and $AB = 0$, then

$$e^{t(A+B)} = e^{tA} + B\int_0^t e^{(t-s)}e^{sA}\,ds, \qquad t \in \mathbb{R}.$$

Alternatively, $e^{t(A+B)} = e^{tA} + B\sum_{n=1}^{\infty}(I + A + \cdots + A^{n-1})\frac{t^n}{n!}, t \in \mathbb{R}$. **Hint:** For the first part, differentiate and use Theorem 13.7. For the second, use induction to show that $(A + B)^n = A^n + B(I + A + \cdots A^{n-1}), n \geq 1$.

Exercise 13.19. ▲ In the set-up of the previous exercise assume that $BA = 0$ and there is a $b \in \mathbb{R}$ such that $B^2 = bB$. Prove that

$$e^{t(A+B)} = e^{tA} + \int_0^t e^{b(t-s)A}e^{sA}B\,ds, \qquad t \in \mathbb{R}.$$

Alternatively, $e^{t(A+B)} = e^{tA} + \sum_{n=1}^{\infty}(b^{n-1} + b^{n-2}A + \cdots + bA^{n-2} + A^{n-1})B\frac{t^n}{n!}, t \in \mathbb{R}$. **Hint:** Proceed as in the previous example using $Be^{tA} = B, t \in \mathbb{R}$.

☞ CHAPTER SUMMARY

The space $\mathcal{L}(\mathbb{X})$ of bounded linear operators mapping a Banach space $\mathbb{X}$ into itself is not only a Banach space but also a Banach algebra with multiplication defined as composition of operators. This provides additional possibilities of manipulating elements of $\mathcal{L}(\mathbb{X})$. In particular, we can use 'power series' of operators to construct inverses of other operators, and thus solve linear equations in $\mathbb{X}$. We can also define exponential functions of bounded linear operators to solve differential equations in $\mathbb{X}$. Again, all of this would be impossible were we not working in a complete space.

14

The Banach–Steinhaus Theorem and Strong Convergence

14.1 Strong Convergence versus Convergence in the Operator Norm

The mode of convergence of bounded linear operators described in the previous chapter, though by all means useful and elegant, is not able to encompass more delicate limit phenomena. For instance, for $t \geq 0$ let's think of the operator $A(t)$ in l^1 given by the formula

$$A(t)\,(\xi_i)_{i\geq 1} = \left(\mathrm{e}^{-it}\xi_i\right)_{i\geq 1}. \tag{14.1}$$

It is easy to see that all $A(t), t \geq 0$ have norms not exceeding 1. It can also be shown that for all $x \in l^1$,

$$\lim_{t\to 0} \|A(t)x - x\| = \lim_{t\to 0} \sum_{i=1}^{\infty} |(\mathrm{e}^{-it} - 1)\xi_i| = 0.$$

This is a direct consequence of the dominated convergence theorem, but can also be obtained in a more elementary way as follows: the sum above equals

$$\sum_{i=1}^{\infty}(1 - \mathrm{e}^{-it})|\xi_i|,$$

and the series $\sum_{i=1}^{\infty} |\xi_i|$ converges. Hence, having given $\epsilon > 0$, we can choose an n so that

$$\sum_{i=n+1}^{\infty} (1 - \mathrm{e}^{-it})|\xi_i| \leq \sum_{i=n+1}^{\infty} |\xi_i| < \frac{\epsilon}{2}.$$

Next, we can choose t so that $\sum_{i=1}^{n}(1 - \mathrm{e}^{-it})|\xi_i| < \frac{\epsilon}{2}$ (since for each $i = 1, \ldots, n$ we have $\lim_{t\to 0} \mathrm{e}^{-it} = 1$). This completes the proof.

As we see, the operators $A(t)$ converge in a natural sense to the identity operator I. Yet, for each $t > 0$,

$$\|A(t)-I\| = \sup_{x\in\ell^1, \|x\|=1} \|A(t)x-x\| \geq \sup_{n\geq 1} \|A(t)e_n-e_n\| = \sup_{n\geq 1}(1-\mathrm{e}^{-nt}) = 1,$$

where the e_n, defined in (10.1), are treated as elements of ℓ^1. This shows that the distance between $A(t)$ and I is constantly 1, and implies that $A(t)$ cannot converge to I in the operator norm. (The same conclusion can be drawn from (12.6) with A replaced by $A(t) - I$.) We have thus encountered an example of convergence in a different, more delicate sense. Here is its formal definition.

14.1 Definition Let $\mathbb{X}$ and $\mathbb{Y}$ be normed linear spaces. We say that a sequence $(A_n)_{n\geq 1}$ of elements of $\mathcal{L}(\mathbb{X}, \mathbb{Y})$ converges to an operator A in the strong topology if for any $x \in \mathbb{X}$ we have $\lim_{n\to\infty} A_n x = Ax$, that is,

$$\lim_{n\to\infty} \|A_n x - Ax\|_{\mathbb{Y}} = 0.$$

We note immediately that convergence in the strong topology is weaker than convergence in the operator norm, which is to say that the latter implies the former. To be more precise: if $(A_n)_{n\geq 1}$ converges in norm to A, then for any $x \in \mathbb{X}$ we have

$$\|A_n x - Ax\| \leq \|A_n - A\| \, \|x\| \underset{n\to\infty}{\longrightarrow} 0,$$

implying convergence of $(A_n)_{n\geq 1}$ in the strong topology. As we have seen above, the converse is not true. It can happen that a sequence $(A_n)_{n\geq 1}$ does not converge in the operator norm to an A

$$\|A_n - A\| \underset{n\to\infty}{\not\to} 0,$$

and yet for all $x \in \mathbb{X}$,

$$\lim_{n\to\infty} \|A_n x - Ax\| = 0. \tag{14.2}$$

In fact, this is not surprising at all, since the operator norm is devised to describe convergence that is uniform with respect to all x in the unit ball – why would all sequences of operators need to converge uniformly? More significantly, convergence in the operator norm, though important and elegant, turns out to happen rather seldom – interesting limit theorems for mathematical analysis and the theory of stochastic processes in particular are too delicate to be discussed in this norm's framework. For instance, in Section 14.3, we will see that the Weierstrass theorem of Chapter 7 provides an example of strong convergence of operators which, however, do not converge in the operator norm, and a similar statement is true about Fejér's theorem of Section 14.4.

14.2 Banach–Steinhaus Theorem

Convergence in the strong topology seems to be a natural notion. However, by looking just at Definition 14.1, we may start to wonder whether it is useful. For, can we be sure that the strong limit of bounded linear operators is bounded? The well-known example of functions $x_n(t) := t^n, t \in [0,1]$ shows that a pointwise limit of continuous functions need not be continuous. Hence, just the mere knowledge of the existence of the limit

$$Ax := \lim_{n\to\infty} A_n x, \qquad x \in \mathbb{X} \tag{14.3}$$

for a sequence $(A_n)_{n\geq 1}$ of bounded linear operators in a Banach space $\mathbb{X}$ should not, and does not, a priori, guarantee that the operator A is bounded. This is troublesome, isn't it?

Hence, the following statement is a kind of mathematical miracle and should be treated with due respect.

> In Banach spaces strong convergence of bounded linear operators implies boundedness of the limit operator.

In particular, it follows that strong convergence is sufficiently user-friendly. Moreover, and this is why this theorem is included in this little book, again it is the completeness that lies behind this curious phenomenon, and leads to far-reaching consequences of the resulting theory.

To understand better the reason why this theorem is true, let's have a closer look at its assumptions. First of all, and this is something we cannot overlook, we have linearity here. The sequence of continuous functions on $[0,1]$ recalled above does not fit into the scenario at hand, because the functions involved are far from being linear. Moreover, the one-dimensional case does not shed more light: it is easy to see that pointwise convergence of the sequence of linear functions[1]

$$f_n(x) = a_n x + b_n, \qquad x \in \mathbb{X},$$

implies convergence of the numerical series $(a_n)_{n\geq 1}$ and $(b_n)_{n\geq 1}$, and this in turn proves uniform convergence in any closed subinterval of $\mathbb{R}$ to the limit linear continuous function. A similar attempt at finding a counterexample in

[1] These functions, even though they *are* called linear, are not linear in the sense used in this book, and hence it would be better to call them *affine*. Nevertheless, they are so close to being linear that they cannot be used as a counterexample to the discussed theorem.

a finite-dimensional space, using matrices, proves to be as futile as the one presented above.

Hence, linearity seems to play a significant role here. But the following simple example shows that linearity alone does not suffice. Namely, let c_{00} be the space of sequences $x = (\xi_i)_{i\geq 1}$ that, starting from a certain coordinate, characteristic for x, not for the space, are composed solely of zeros. (As usual, we equip c_{00} with the norm $\|x\| = \max_{i\geq 1} |\xi_i|$.) We consider the sequence of functionals on c_{00} defined by

$$A_n (\xi_i)_{i\geq 1} = \sum_{i=1}^{n} \xi_i \in \mathbb{R}, \qquad n \geq 1. \tag{14.4}$$

This sequence converges in the strong topology to the functional A defined by

$$A (\xi_i)_{i\geq 1} = \sum_{i=1}^{\infty} \xi_i .$$

In fact, the sum on the right-hand side is composed of finitely many non-zero elements and thus we have $A_n x = Ax$ for all $n \geq n_0$, where n_0 depends on x. However, A is not bounded (and thus not continuous). Indeed, for $x_n := \sum_{i=1}^{n} e_i$, where the e_n defined in (10.1) are thought of as elements of c_{00}, we have $\|x_n\| = 1$ and at the same time $Ax_n = n$, implying $\sup_{\|x\|=1} |Ax| = \infty$. Here, the framed statement does not apply because c_{00} is not a Banach space; c_{00} is not complete.

We conclude thus that the statement involves a unique blend of intertwined assumptions (see also Exercise 14.5). As we shall see next, it is a direct consequence of the following theorem of S. Banach and H. Steinhaus; this theorem is also known as the principle of uniform boundedness.

14.2 Theorem (Banach–Steinhaus) *Suppose that bounded linear operators A_n, $n \geq 1$ are defined in a Banach space and have values in a normed linear space (the latter space need not be complete). If for all $x \in \mathbb{X}$,*

$$\sup_{n\geq 1} \|A_n x\| < \infty, \tag{14.5}$$

then

$$\sup_{n\geq 1} \|A_n\| < \infty. \tag{14.6}$$

We stress that the symbol $\|\cdot\|$ in (14.5) denotes the norm in the vector space $\mathbb{Y}$, whereas in (14.6) it denotes the operator norm of $\mathcal{L}(\mathbb{X}, \mathbb{Y})$.

The most popular proofs of this theorem, to be found in many monographs, including e.g., [11, 13, 31, 33, 35], are quite involved and are based either on the Baire category theorem or the so-called gliding hump argument (see also Chapter 15). The following elementary argument is due to A.D. Sokal [38].

14.3 Lemma *If A is a bounded linear operator, then for all $x \in \mathbb{X}$ and $r > 0$ we have*

$$r\|A\| \le \sup_{y \in B(x,r)} \|Ay\|,$$

where $B(x,r)$ denotes the closed ball centered at x with radius r, that is, the set of $y \in \mathbb{X}$ such that $\|x - y\| \le r$.

Proof If $\|z\| \le 1$, then $\|rz\| \le r$ and we have

$$2rz = rz - x + x + rz,$$

which implies in turn that

$$r\|Az\| \le \tfrac{1}{2}\| - A(x - rz)\| + \tfrac{1}{2}\|A(x + rz)\|.$$

Both $x - rz$ and $x + rz$ belong to $B(x,r)$; hence the right-hand side does not exceed $\sup_{y \in K(x,r)} \|Ay\|$. Taking the supremum over $\|z\| \le 1$, we obtain the claim. □

14.4 Corollary *Regardless of the choice of x and $r > 0$, in the closed ball centered at x and with radius r, a y can be found such that $\|Ay\| \ge \frac{2}{3}r\|A\|$.*

Proof of the Banach–Steinhaus Theorem Assume the thesis does not hold. Then, for each n we can find an $m = m(n)$ such that $\|A_m\| \ge 4^n$. To simplify notations, in what follows we assume, without loss of generality, that $\|A_n\| \ge 4^n$ – in the general case, it suffices to consider $B_n := A_{m(n)}$.

The proof hinges on the following inductive construction of a sequence $(x_n)_{n \ge 0}$ in $\mathbb{X}$: we put $x_0 = 0$ and, having constructed x_{n-1}, we define x_n as an element of $B(x_{n-1}, \frac{1}{3^n})$ such that

$$\|A_n x_n\| \ge \tfrac{2}{3}\tfrac{1}{3^n}\|A_n\|;$$

such an x_n exists by Corollary 14.4.

I claim that $(x_n)_{n \ge 1}$ is a Cauchy sequence. Indeed, for $m > n$,

$$\begin{aligned}\|x_m - x_n\| &\le \|x_m - x_{m-1}\| + \cdots + \|x_{n+1} - x_n\| \\ &\le \tfrac{1}{3^m} + \cdots + \tfrac{1}{3^{n+1}} < \frac{\frac{1}{3^{n+1}}}{1 - \frac{1}{3}} = \tfrac{1}{2}\tfrac{1}{3^n}.\end{aligned}$$

Therefore, $\mathbb{X}$ being a Banach space, there exists the limit $x = \lim_{n\to\infty} x_n$, and we have furthermore $\|x - x_n\| \leq \frac{1}{2}\frac{1}{3^n}$. Hence,

$$\begin{aligned}\|A_n x\| &\geq \|A_n x_n\| - \|A_n(x - x_n)\| \\ &\geq \tfrac{2}{3}\tfrac{1}{3^n}\|A_n\| - \tfrac{1}{2}\tfrac{1}{3^n}\|A_n\| = \tfrac{1}{6}\tfrac{1}{3^n}\|A_n\|\end{aligned}$$

because we know that $\|A_n(x - x_n)\| \leq \frac{1}{2}\frac{1}{3^n}\|A_n\|$. This proves, however, that

$$\|A_n x\| \geq \tfrac{1}{6}\left(\tfrac{4}{3}\right)^n,$$

contradicting (14.5). This contradiction completes the proof. □

It is now easy to prove the framed statement. We begin by noting that A defined by

$$Ax := \lim_{n\to\infty} A_n x \tag{14.7}$$

is obviously linear, for using the linearity of A_n's and of the operation of taking the limit, we can write

$$\begin{aligned}A(\alpha x + \beta y) &= \lim_{n\to\infty} A_n(\alpha x + \beta y) = \lim_{n\to\infty} (\alpha A_n x + \beta A_n y) \\ &= \alpha \lim_{n\to\infty} A_n x + \beta \lim_{n\to\infty} A_n y = \alpha Ax + \beta Ay.\end{aligned}$$

To prove boundedness, in turn, we reason as follows: the inequality

$$|\|A_n x\| - \|Ax\|| \leq \|A_n x - Ax\|$$

shows that convergence of the sequence $(A_n x)_{n\geq 1}$ implies convergence of the numerical sequence $(\|A_n x\|)_{n\geq 1}$ (to $\|Ax\|$). However, a convergent sequence is bounded. Hence, existence of the limit (14.7) implies (14.5), and thus, by the Banach–Steinhaus Theorem, $M = \sup_{n\geq 1}\|A_n\|$ is finite. For this M and any $x \in \mathbb{X}$ we have furthermore

$$\|Ax\| = \lim_{n\to\infty} \|A_n x\| \leq \limsup_{n\to\infty} \|A_n\|\,\|x\| \leq M\|x\|,$$

completing the proof.

14.3 Bernstein Polynomials

The proof of the Weierstrass theorem presented in Chapter 7 is not constructive, as it does not provide an efficient way to approximate a function by a sequence of polynomials. A constructive proof was given by Sergey Bernstein in 1912. He showed that, for each $x \in C[0, 1]$, the polynomials

$$w_{n,x}(t) = \sum_{k=0}^{n} \binom{n}{k} x\,(k/n)\, t^k (1-t)^{n-k},$$

which are built by means of x, converge to x as $n \to \infty$. They are now known as Bernstein polynomials.

In this section, we will look at this result from the perspective of convergence of operators. Namely, we define the operators A_n in $C[0,1]$ by the formula

$$A_n x(t) = w_{n,x}(t) = \sum_{k=0}^{n} x\,(k/n) \binom{n}{k} t^k (1-t)^{n-k},$$

so that A_n corresponds to the nth Bernstein polynomial, and are interested in the limit of A_n. It is easy to see that $A_n, n \geq 1$ are bounded with bound equal to 1:

$$\begin{aligned} \|A_n x\| &= \sup_{t\in[0,1]} |A_n x(t)| \leq \sup_{x\in[0,1]} \sum_{k=0}^{n} |x\,(k/n)| \binom{n}{k} t^k (1-t)^{n-k} \\ &\leq \|x\| \sup_{t\in[0,1]} \sum_{k=0}^{n} \binom{n}{k} t^k (1-t)^{n-k} = \|x\| \end{aligned}$$

by Newton's binomial formula. Moreover, if $x_0(t) = 1$ for all $t \in [0,1]$, then $Ax_0 = x_0$, and obviously $\|x_0\| = 1$, proving that not only do we have $\|A_n\| \leq 1$ but also $\|A_n\| = 1, n \geq 1$.

We will show that, for each x,

$$\lim_{n\to\infty} \|A_n x - x\| = 0, \tag{14.8}$$

which is just a restatement of the Bernstein version of the Weierstrass theorem, whereas, for all n,

$$\|A_n - I\| \geq 1. \tag{14.9}$$

The last formula excludes the possibility of convergence of $(A_n)_{n\geq 1}$ in the operator norm. Let's explain this in more detail: equation (14.8), once proved, shows that there can be no limit of $(A_n)_{n\geq 1}$ in the operator norm, other than the identity operator I, because we have already shown that convergence in the operator norm implies strong convergence. On the other hand, in view of (14.9), I cannot be the limit of A_n in the operator norm. Hence, $(A_n)_{n\geq 1}$ has no limit in the operator norm whatsoever.

Hence, a classical theorem of mathematical analysis, a cornerstone for many arguments, cannot be described in terms of convergence in the operator norm: the phenomenon is more delicate, more subtle, and requires thinking in terms of strong convergence.

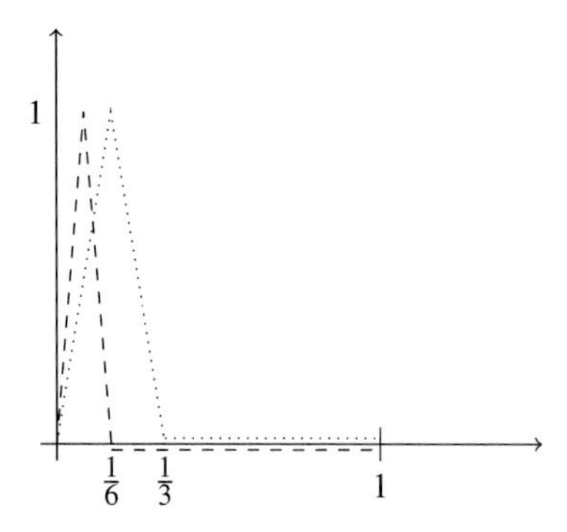

Figure 14.1 Functions x_3 (dotted line) and x_6 (dashed line); in the interval $[\frac{1}{3}, 1]$ they coincide.

The reader has probably already observed that Bernstein polynomials are related to the binomial distribution of probability theory. To recall, a random variable X is said to be binomially distributed with parameters n and $t \in [0, 1]$ if

$$\Pr(X = k) = \binom{n}{k} t^k (1 - t)^{n-k} \qquad \text{for} \qquad k = 0, \ldots, n.$$

Such a random variable describes the number of successes in n independent Bernoulli trials, provided that $t \in [0, 1]$ is the probability of success in one trial. In these terms we can write

$$A_n x(t) = E\, x(n^{-1} X_n),$$

where E denotes expected value, and X_n is a binomial variable with parameters n and t.

We also recall that the expected value of a binomial variable with parameters n and t is nt, and its variance is $nt(1 - t)$, see [20]. For the proof of (14.8) we also need the following simple inequality of Chebyshev: if X is a random variable with expected value μ and variance σ^2, then for any positive δ, we have

$$\Pr(|X - \mu| > \delta) \le \frac{\sigma^2}{\delta^2}. \tag{14.10}$$

Let's prove (14.9) first: let $x_n(t) = 0 \vee (1 - |2nt - 1|)$ for $t \in [0, 1]$, see Figure 14.1. Clearly, $\|x_n\| = 1$ (note $x(1/(2n)) = 1$), but $x_n(j/n) = 0$, for $j = 0, 1, \ldots, n$, proving that $A_n x_n = 0$. Hence, $\|A_n - I\| \ge \|A_n x_n - x_n\| = \|x_n\| = 1$. This completes the proof of (14.9).

Turning to the proof of (14.8), we note that x, being continuous on a closed interval, is uniformly continuous. Hence, for any $\epsilon > 0$, a $\delta > 0$ can be chosen so that $|x(s) - x(t)| < \frac{\epsilon}{2}$ as long as $s, t \in [0, 1]$ satisfy $|s - t| < \delta$. Since

$$\sum_{k=0}^{n} \binom{n}{k} t^k (1-t)^{n-k} = 1,$$

we also have

$$x(t) = \sum_{k=0}^{n} x(t) \binom{n}{k} t^k (1-t)^{n-k}.$$

Hence, the difference between $x(t)$ and $A_n x(t)$ is no larger than

$$\sum_{k=0}^{n} |x(t) - x(k/n)| \binom{n}{k} t^k (1-t)^{n-k}.$$

Let's split this expression into two sums. The first of these, say Σ_1, includes all terms with indices such that $|\frac{k}{n} - t| \geq \delta$; the other, say Σ_2, contains all the remaining terms. Because of the choice of δ, Σ_2 does not exceed $\frac{\epsilon}{2}$ times the probability that $|\frac{k}{n} - t| < \delta$. Since this probability, obviously, cannot exceed 1, we have $\Sigma_2 \leq \frac{\epsilon}{2}$. On the other hand, in the first sum, the numbers $|x(t) - x(k/n)|$ are no larger than $2\|x\|$, whereas the corresponding probabilities add up to $\Pr(|n^{-1}X_n - t| \geq \delta)$, which, in turn, by Chebyshev's inequality, does not exceed $\frac{t(1-t)}{n\delta^2}$. Therefore, $\Sigma_1 \leq 2\|x\|\frac{t(1-t)}{n\delta^2}$. Since the maximum of the function $[0,1] \ni t \mapsto t(1-t)$ is attained at $t = \frac{1}{2}$, and equals $\frac{1}{4}$, we see that $\Sigma_2 \leq \frac{\|x\|}{2n\delta^2}$. This shows that, as long as $n \geq \frac{\|x\|}{\epsilon\delta^2}$, the norm of $x_n - x$ does not exceed $\frac{\epsilon}{2} + \frac{\epsilon}{2} = \epsilon$, as claimed.

14.4 Fejér's Theorem

The theorem of Fejér[2] says that a continuous function $x\colon [-\pi,\pi] \to \mathbb{R}$ such that $x(-\pi) = x(\pi)$ can be uniformly approximated by trigonometric polynomials, that is, by finite linear combinations of functions $\sin nt$ and $\cos nt$ with n taking on the values of one or more natural numbers. We have already obtained this result as Corollary 7.8, but it is worth stressing that, just as in Section 14.3, the proof based on the Stone–Weierstrass theorem does not provide an explicit formula for the approximating polynomials. In this section, we will show that the specific sequence of trigonometric polynomials considered by Fejér provides another example of strong convergence of operators.

[2] Lipót Fejér (1880–1959) was a Hungarian mathematician; he was only 19 when he proved his now-famous theorem.

As a preparation, we recall that convergence of a sequence $(x_n)_{n\geq 1}$ of elements of a Banach space implies convergence of $\left(\frac{x_1+\cdots+x_n}{n}\right)_{n\geq 1}$ to the same limit, but not vice versa. This means that $\left(\frac{x_1+\cdots+x_n}{n}\right)_{n\geq 1}$ converges more often than $(x_n)_{n\geq 1}$ itself, that is, that the former sequence is more amenable to convergence than the latter. Fejér's idea was therefore as follows: instead of attempting to prove that the sequence $(s_n)_{n\geq 1}$ of trigonometric polynomials, defined by

$$s_n(t) := \frac{a_0}{2} + \sum_{\ell=1}^{n} (a_\ell \cos \ell t + b_\ell \sin \ell t), \tag{14.11}$$

where (see (10.7))

$$a_\ell := \frac{1}{\pi} \int_{-\pi}^{\pi} x(s) \cos \ell s \, \mathrm{d}s \qquad \text{and} \qquad b_\ell := \frac{1}{\pi} \int_{-\pi}^{\pi} x(s) \sin \ell s \, \mathrm{d}s \tag{14.12}$$

converges uniformly to x, let's prove that so does $(\sigma_n)_{n\geq 1}$ where

$$\sigma_n = \frac{s_0 + \cdots + s_{n-1}}{n}, \qquad n \geq 1. \tag{14.13}$$

In what follows we will see that this idea works well: indeed, $(\sigma_n)_{n\geq 1}$ behaves better than $(s_n)_{n\geq 1}$.

14.4.1 Trigonometric Identities

For the proof of Fejér's theorem we need the following well-known identities:

$$\cos\alpha \cos\beta + \sin\alpha \sin\beta = \cos(\alpha - \beta), \tag{14.14}$$

$$\sin\alpha - \sin\beta = 2\cos\frac{\alpha+\beta}{2} \sin\frac{\alpha-\beta}{2}, \tag{14.15}$$

$$\cos\alpha - \cos\beta = 2\sin\frac{\alpha+\beta}{2} \sin\frac{\beta-\alpha}{2}, \qquad \alpha, \beta \in \mathbb{R}. \tag{14.16}$$

It is probably easiest to check them using Euler's identities (7.4).

14.5 Lemma *If $\alpha \in \mathbb{R}$ is not of the form $\alpha = 2\ell\pi$ for some integer ℓ,*

$$k_n(\alpha) := \tfrac{1}{2} + \cos\alpha + \cdots + \cos n\alpha = \frac{\sin(n+\frac{1}{2})\alpha}{2\sin\frac{\alpha}{2}} = \frac{\cos n\alpha - \cos(n+1)\alpha}{2(1-\cos\alpha)}, \qquad n \geq 0. \tag{14.17}$$

Otherwise, $k_n(\alpha) = n + \frac{1}{2}$.

Proof By (14.15),

$$
\begin{aligned}
2k_n(\alpha)\sin\frac{\alpha}{2} &:= \sin\frac{\alpha}{2} + 2\cos\alpha\sin\frac{\alpha}{2} + \cdots + 2\cos n\alpha\sin\frac{\alpha}{2} \\
&= \sin\frac{\alpha}{2} + (\sin\frac{3\alpha}{2} - \sin\frac{\alpha}{2}) + \cdots + (\sin(n+\tfrac{1}{2})\alpha - \sin(n-\tfrac{1}{2})\alpha) \\
&= \sin(n+\tfrac{1}{2})\alpha,
\end{aligned}
$$

proving the first part of (14.17). On the other hand, by (14.16),

$$\cos n\alpha - \cos(n+1)\alpha = 2\sin(n+\tfrac{1}{2})\sin\frac{\alpha}{2}.$$

This shows the other equality in (14.17) because $1-\cos\alpha = 2\sin^2\frac{\alpha}{2}$. The rest is clear. □

Here is an immediate consequence of our lemma.

14.6 Corollary *For α not of the form $\alpha = 2\ell\pi$,*

$$\kappa_n(\alpha) := \frac{k_0(\alpha)+\cdots+k_{n-1}(\alpha)}{n} = \frac{1-\cos n\alpha}{2n(1-\cos\alpha)}, \qquad n \ge 1. \qquad (14.18)$$

Otherwise, $\kappa_n(\alpha) = \frac{n}{2}$.

Before proceeding, we note that each $k_n, n \ge 0$, being a sum of continuous functions of period 2π, is itself a continuous function of period 2π, and the same is true of $\kappa_n, n \ge 1$. It can be checked directly, for example, that $\lim_{\alpha\to 0} k_n(\alpha) = k_n(0)$ and $\lim_{\alpha\to 0}\kappa_n(\alpha) = \kappa_n(0)$.

14.4.2 Back to the Fourier Series

Let's come back to s_n and σ_n defined at the beginning of this section. By (14.11)–(14.12) and (14.14), we can write

$$
\begin{aligned}
s_n(t) &= \frac{1}{2\pi}\int_{-\pi}^{\pi} x(t)\,\mathrm{d}t \\
&\quad + \sum_{\ell=1}^{n}\left(\frac{1}{\pi}\int_{-\pi}^{\pi} x(s)\cos\ell s\,\mathrm{d}s\cos\ell t + \frac{1}{\pi}\int_{-\pi}^{\pi} x(s)\sin\ell s\,\mathrm{d}s\sin\ell t\right) \\
&= \frac{1}{\pi}\int_{-\pi}^{\pi} x(s)\left[\frac{1}{2} + \sum_{\ell=1}^{n}(\cos\ell s\cos\ell t + \sin\ell s\sin\ell t)\right]\mathrm{d}s \\
&= \frac{1}{\pi}\int_{-\pi}^{\pi} x(s)\left[\frac{1}{2} + \sum_{\ell=1}^{n}\cos\ell(t-s)\right]\mathrm{d}s \\
&= \frac{1}{\pi}\int_{-\pi}^{\pi} x(s)k_n(t-s)\,\mathrm{d}s. \qquad (14.19)
\end{aligned}
$$

The second to last equality here is a direct consequence of (14.11), and k_n is defined in (14.17). Therefore, by Corollary 14.6,

$$\sigma_n(t) = \frac{1}{\pi}\int_{-\pi}^{\pi} x(s)\kappa_n(t-s)\,\mathrm{d}s, \qquad t \in [-\pi,\pi], \tag{14.20}$$

and this is a key relation to be used later in its various forms.

To proceed, it will be convenient to think of our x as a 2π-periodic function defined on the entire real line. In other words, we identify the $x\colon [-\pi,\pi] \to \mathbb{R}$ with its extension given by

$$x(t + 2\ell\pi) = x(t), \qquad t \in [-\pi,\pi], \ell \in \mathbb{Z}.$$

Changing variables, and noting that for any $t \in \mathbb{R}$ and 2π-periodic continuous function $y\colon \mathbb{R} \to \mathbb{R}$ (see Exercise 14.8),

$$\int_{t-\pi}^{t+\pi} y(s)\,\mathrm{d}s = \int_{-\pi}^{\pi} y(s)\,\mathrm{d}s, \tag{14.21}$$

we obtain the following incarnation of (14.20):

$$\sigma_n(t) = \frac{1}{\pi}\int_{-\pi}^{\pi} x(t-s)\kappa_n(s)\,\mathrm{d}s = \frac{1}{\pi}\int_{-\pi}^{\pi} x(t+s)\kappa_n(s)\,\mathrm{d}s, \qquad t \in [-\pi,\pi], \tag{14.22}$$

where the second equality holds because $s \mapsto \kappa_n(s)$ is even (by (14.18)).

14.4.3 Fejér's Theorem in the Language of Operators

Fejér's theorem says that the sequence $(\sigma_n)_{n\geq 1}$ of trigonometric polynomials formed by means of x converges uniformly to x. Let's restate this in the language of convergence of operators. We will work in the space of $C_{\mathrm{p}}[-\pi,\pi]$ of continuous functions $x\colon [-\pi,\pi] \to \mathbb{R}$ such that $x(-\pi) = x(\pi)$ (the subscript 'p' stands for 'periodic'). When equipped with the maximum norm, $C_{\mathrm{p}}[-\pi,\pi]$ is a Banach space. We define bounded linear operators $A_n, n \geq 1$ in $C_{\mathrm{p}}[-\pi,\pi]$ as follows:

$$A_n x(t) = \sigma_n(t) = \frac{1}{\pi}\int_{-\pi}^{\pi} x(s+t)\kappa_n(s)\,\mathrm{d}s, \qquad x \in C_{\mathrm{p}}[-\pi,\pi], t \in [-\pi,\pi]. \tag{14.23}$$

The reader will have no difficulty in checking that A_n maps the space $C_{\mathrm{p}}[-\pi,\pi]$ into $C_{\mathrm{p}}[-\pi,\pi]$.

14.7 Theorem (Fejér) *The operators $A_n, n \geq 1$ converge strongly to the identity operator in $C_{\mathrm{p}}[-\pi,\pi]$.*

This obviously means that

$$\max_{t\in[-\pi,\pi]} |x(t) - \sigma_n(t)| = \max_{t\in[-\pi,\pi]} |x(t) - A_n x(t)| \underset{n\to\infty}{\longrightarrow} 0.$$

We prove this theorem in Section 14.4.5.

14.4.4 Probabilistic Insight

The fact that, by Corollary 14.6,

$$\kappa_n(s) \geq 0, \qquad s \in \mathbb{R}, n \geq 1, \tag{14.24}$$

is of crucial importance for the proof of Fejér's theorem. We note that $k_n, n \geq 0$ *do not* possess this property. By taking the average we indeed obtained a nicer object. Moreover, since $\int_{-\pi}^{\pi} \cos ns \,\mathrm{d}s = 0$ for $n \geq 1$, we have, by definition (14.17), $\int_{-\pi}^{\pi} k_n(s)\,\mathrm{d}s = \pi, n \geq 0$. If follows that

$$\frac{1}{\pi}\int_{-\pi}^{\pi} \kappa_n(s)\,\mathrm{d}s = 1, \qquad n \geq 1. \tag{14.25}$$

Properties (14.24)–(14.25) combined allow us to think of $\frac{1}{\pi}\kappa_n$ as the density of a random variable, say X_n, taking values in $[-\pi,\pi]$. In other words,

$$\Pr(X_n \in [a,b]) = \frac{1}{\pi}\int_a^b \kappa_n(s)\,\mathrm{d}s$$

for $a, b \in [-\pi,\pi]$.

14.8 Lemma *Random variables $X_n, n \geq 1$ are centered (i.e., $E\,X_n = 0$) and there exists a constant M such that*

$$E\,X_n^2 \leq \frac{M}{n}, \qquad n \geq 1.$$

Proof Fix $n \geq 1$. The variable X_n is bounded, because it takes values in $[-\pi,\pi]$. Hence, its first moment $E\,X_n$ exists. Moreover, by (14.18), κ_n is even. Therefore, $s \mapsto s\kappa_n(s)$ is odd, and we obtain $E\,X_n = \frac{1}{\pi}\int_{-\pi}^{\pi} s\kappa_n(s)\,\mathrm{d}s = 0$. Next, again by (14.18),

$$s^2\kappa_n(s) \leq \frac{s^2}{2n(1-\cos s)}, \qquad s \in [-\pi,\pi] \setminus \{0\}.$$

Since $([-\pi,\pi] \setminus \{0\}) \ni s \mapsto \frac{s^2}{1-\cos s}$ is continuous and $\lim_{s\to 0} \frac{s^2}{1-\cos s} = 2$, there is an M such that $\sup_{s\in[-\pi,\pi]} s^2\kappa_n(s) \leq \frac{M}{2n}$. It follows that

$$E\,X_n^2 = \frac{1}{\pi}\int_{-\pi}^{\pi} s^2\kappa_n(s)\,\mathrm{d}s \le \frac{M}{n},$$

as claimed. □

14.4.5 Proof of Fejér's Theorem

With Lemma 14.8 under our belt, we boldly proceed to the proof of Fejér's theorem. We start by writing, for $t \in [-\pi,\pi], n \ge 1$ and $x \in C_{\mathrm{p}}[-\pi,\pi]$,

$$\begin{aligned}|A_n x(t) - x(t)| &= \left|\frac{1}{\pi}\int_{-\pi}^{\pi} x(t+s)\kappa_n(s) - x(t)\frac{1}{\pi}\int_{-\pi}^{\pi}\kappa_n(s)\,\mathrm{d}s\right| \\ &\le \frac{1}{\pi}\int_{-\pi}^{\pi}|x(t+s) - x(t)|\kappa_n(s)\,\mathrm{d}s.\end{aligned}$$

Next, for a small $\delta > 0$ to be chosen later, we split the last integral into two, say I_1 and I_2; in the first we integrate over $[-\pi, -\pi] \setminus (-\delta,\delta)$ and in the second, over $(-\delta,\delta)$. Then

$$I_1 \le 2\|x\|\frac{1}{\pi}\int_{[-\pi,-\pi]\setminus(-\delta,\delta)}\kappa_n(s)\,\mathrm{d}s = 2\|x\|\,\Pr(|X_n| \ge \delta)$$

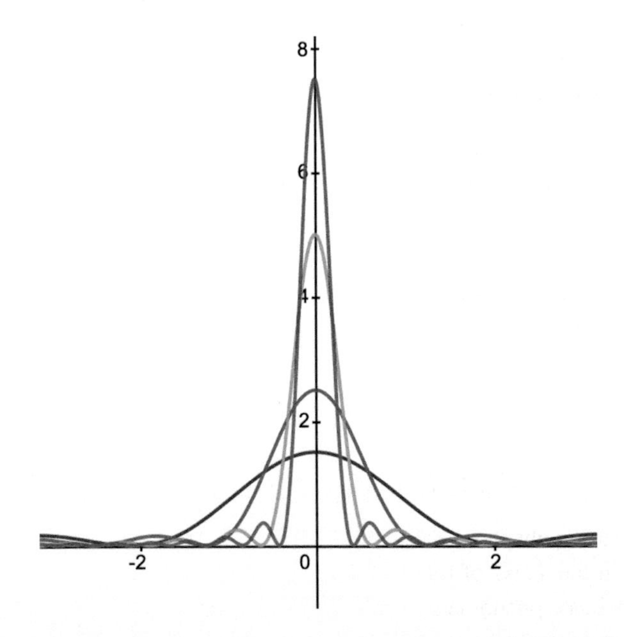

Figure 14.2 Functions κ_n in the interval $[-\pi,\pi]$ for $n = 3, 5, 10$ and 15. As n increases, so does the value of $\kappa_n(0)$, and more and more of the area under the graph of κ_n concentrates in small intervals around 0. The total area under each graph does not depend on n and equals π.

and

$$I_2 \le \max_{|s|\le\delta} |x(t+s) - x(t)|$$

because $\frac{1}{\pi}\int_{-\delta}^{\delta} \kappa_n(s)\,\mathrm{d}s < \frac{1}{\pi}\int_{-\pi}^{\pi} \kappa_n(s)\,\mathrm{d}s = 1$. It follows that

$$\begin{aligned}\|A_n x - x\| &= \max_{t\in[-\pi,\pi]} |A_n x(t) - x(t)| \\ &\le 2\|x\| \Pr(|X_n| \ge \delta) + \sup_{|s|\le\delta, t\in[-\pi,\pi]} |x(t+s) - x(s)|.\end{aligned}$$

So prepared and given $\epsilon > 0$, we first choose a $\delta > 0$ so that $|x(t+s)-x(t)| < \frac{\epsilon}{2}$ provided that $|s| \le \delta$ for all $t \in \mathbb{R}$; for such a δ the second term above does not exceed $\frac{\epsilon}{2}$. Such a choice is possible because $[-\pi,\pi] \ni s \mapsto x(s)$ is uniformly continuous, and its 2π-periodic extension shares this property. Also, by Chebyshev's inequality (14.10) (by Lemma 14.8 we have $\mu = 0$ and $\sigma^2 \le \frac{M}{n}$),

$$\Pr(|X_n| \ge \delta) \le \frac{M}{n\delta^2}.$$

Therefore, as long as $n > \frac{4M\|x\|}{\epsilon\delta^2}$, the second term is smaller than $\frac{\epsilon}{2}$. This shows that for so-defined δ and n, $\|A_n x - x\| < \epsilon$, completing the proof.

14.9 Remark As in the case of Bernstein polynomials, $(A_n)_{n\ge1}$ does not converge to identity in the operator norm. To see this, note that for $x = x_n$ given by $x(t) = x_n(t) = \frac{1}{\sqrt{\pi}} \cos nt, t \in [-\pi,\pi]$, all the Fourier coefficients a_ℓ, b_ℓ of (14.12) with indices $\ell \le n-1$ are zero. Hence, $s_\ell, \ell \le n-1$ are also zero, and so is $A_n x_n = \sigma_n$. Since $\|x_n\| = 1$, we obtain thus $\|A_n - I\| \ge \|A_n x_n - x_n\| = \|x_n\| = 1$. This excludes the possibility of $\lim_{n\to\infty} \|A_n - I\| = 0$. An intuition as to how κ_n's behave can be gained by looking at Figure 14.2.

14.5 Poisson Approximation

In the foregoing chapters I was very bold in declaring that the convergence of operators in the sense of the operator norm is too crude a tool to be able to describe delicate phenomena. And I was right. But it would be unfair to leave the impression that there are no interesting results we can prove with its help; this is simply untrue. In this section, we will show that the classical Poisson approximation of probability theory involves convergence in operator norm.

To recall, the theorem this section is devoted to can be phrased as follows.

Suppose $X_n, n \geq 1$ are binomial variables; X_n has parameters n and p_n, and these parameters are chosen in such a way that $\lim_{n\to\infty} np_n = \lambda > 0$. Then

$$\lim_{n\to\infty} \Pr(X_n = k) = e^{-\lambda} \frac{\lambda^k}{k!}. \tag{14.26}$$

(The probabilities on the right-hand side are those of the Poisson distribution with parameter λ.)

To rewrite this result in the language of operators, let's imagine that initially we have a certain number of dollars, and that we can build our wealth by participating in Bernoulli trials – each success increases the amount we possess by \$1, and a failure does not hurt us at all. In such a case, if $(\xi_i)_{i\geq 0}$ is the distribution of the amount we possess initially, so that

$$\xi_i = \Pr(\text{initial amount} = i),$$

then

$$\xi_i' := \Pr(\text{new amount} = i) = p\xi_{i-1} + (1-p)\xi_i \qquad (\text{we agree that } \xi_{-1} = 0).$$

Since the manipulations performed above on the distribution make sense also for all absolutely summable sequences $(\xi_i)_{i\geq 0}$, we can consider the operator A in the space ℓ^1, defined by

$$A\,(\xi_i)_{i\geq 0} := (p\xi_{i-1} + (1-p)\xi_i)_{i\geq 0}, \qquad (\xi_i)_{i\geq 0} \in \ell^1. \tag{14.27}$$

Clearly, A is bounded with norm one.

Coming back to the probabilistic interpretation, we note that, if $(\xi_i)_{i\geq 0}$ is the distribution of our initial amount, then $A^2\,(\xi_i)_{i\geq 0}$ is the distribution after two independent Bernoulli trials, and, more generally, $A^n(\xi_i)_{i\geq 1}$ is the distribution after n independent Bernoulli trials. It follows that the operator A^n describes the situation in which our initial amount increases by a binomially distributed random variable with parameters n and p.

Hence, our goal is to find the limit of the sequence $\left(A_n^n\right)_{n\geq 1}$, for A_n defined by (14.27) with p replaced by p_n. We aim to show that

$$\lim_{n\to\infty} A_n^n = P_\lambda, \tag{14.28}$$

where P_λ is the operator related to the Poisson distribution with parameter λ, that is, describing the situation in which our amount increases by a Poisson distributed number of dollars.

Such an increase results in the total of i dollars in $i+1$ possible ways. Either we had i dollars from the start, and the Poisson variable increased our capital by 0, or we had $i-1$ dollars, and the Poisson variable was 1, and so on; the last possibility is that we had 0 dollars and the random variable was i. Denoting by λ_k the probability on the right-hand side of (14.26) we obtain thus the following formula for the probability ξ_i' of possessing i dollars after the random experiment:

$$\xi_i' = \xi_i\lambda_0 + \xi_{i-1}\lambda_1 + \cdots + \xi_0\lambda_i.$$

This means that

$$P_\lambda\,(\xi_i)_{i\geq 0} = \left(\sum_{j=0}^{i} \xi_{i-j}\lambda_i\right)_{i\geq 0}.$$

Proof of (14.28) The following representation of P_λ by means of the exponential function of the shift operator is a key to the proof:

$$P_\lambda = \mathrm{e}^{\lambda(R-I)}. \tag{14.29}$$

The operator R is given here by $R\,(\xi_i)_{i\geq 0} = (\xi_{i-1})_{i\geq 0}$ (as before, $\xi_{-1} = 0$) and describes the deterministic increase of our wealth by one dollar.

By Exercise 13.12, (14.29) is equivalent to

$$P_\lambda = \mathrm{e}^{-\lambda} \sum_{k=0}^{\infty} \frac{\lambda^k}{k!} R^k = \sum_{k=0}^{\infty} \lambda_k R^k,$$

and the latter formula can be interpreted as follows: operator P_λ shifts $(\xi_i)_{i\geq 0}$ by k units to the right with probability λ_k. Since this is an equivalent definition of P_λ, (14.29) is established.

This formula has two immediate consequences. First of all,

$$P_\lambda = \mathrm{e}^{\lambda(R-I)} = \mathrm{e}^{n\frac{\lambda}{n}(R-I)} = \left[\mathrm{e}^{\frac{\lambda}{n}(R-I)}\right]^n = P_{\frac{\lambda}{n}}^n.$$

Furthermore, using $\|R\| = 1$, we can prove that $\|P_\lambda\| = 1$.

Next, we note the following inequality: if $B_1, \ldots, B_n$ and $C_1, \ldots, C_n$ are contractions in a Banach space $\mathbb{X}$, then (see Exercise 14.9)

$$\|B_n B_{n-1} \cdots B_1 - C_n C_{n-1} \cdots C_1\| \leq \sum_{i=1}^{n} \|B_i - C_i\|. \tag{14.30}$$

In particular, for any contractions B and C,

$$\|B^n - C^n\| \leq n\|B - C\|.$$

The rest is now simple. Since

$$\begin{aligned}\|A_n^n - P_\lambda\| = \|A_n^n - P_{\frac{\lambda}{n}}^n\| &\le n\|A_n - P_{\frac{\lambda}{n}}\| \\ &\le \|n(A_n - I) - n(P_{\frac{\lambda}{n}} - I)\|,\end{aligned}$$

it suffices to show that the sequences $(n(A_n - I))_{n\ge1}$ and $\left(n(P_{\frac{\lambda}{n}} - I)\right)_{n\ge1}$ converge to the same limit in the operator norm. To this end, using the second part of Lemma 13.6, we establish first that

$$\lim_{n\to\infty} n(P_{\frac{\lambda}{n}} - I) = \lambda \lim_{n\to\infty} \frac{n}{\lambda}\left(\mathrm{e}^{\frac{\lambda}{n}(R-I)} - I\right) = \lambda \lim_{t\to0} \frac{1}{t}\left(\mathrm{e}^{t(R-I)} - I\right) = \lambda(R - I).$$

Next, by definition of A_n, we have $A_n = p_n R + (1 - p_n)I$, and this shows that

$$n(A_n - I) = np_n(R - I) \longrightarrow \lambda(R - I),$$

completing the proof. □

Finally, let's comment on the relation between (14.28) and (14.26). To this end, we think of the vector $e_0 = (1, 0, 0, \dots) \in \ell^1$ with all but the first coordinate equal to 0. This vector describes the situation in which our initial capital is deterministically 0. Then $A_n^n e_0$ can be identified with the binomial distribution: after the random experiment we possess as many dollars as there were successes in the n independent Bernoulli trials. Now, formula (14.28) tells us that $A_n^n e_0$ tends to $P_\lambda e_0$, which represents the Poisson distribution. Since convergence in the norm of ℓ^1 implies convergence of coordinates, (14.26) follows. In other words, (14.28) indeed implies (14.26).

But the former is in fact much stronger than the latter: as already mentioned, the latter speaks of convergence of coordinates whereas the former of at least convergence in the norm of ℓ^1. It can be argued that using Scheffé's theorem[3] one can deduce convergence of $\left(A_n^n\right)_{n\ge1}$ in the strong topology from (14.26). But our result is in fact yet stronger than this: we have proved that the operators A_n^n converge to P_λ in the operator norm, that is, uniformly with respect to $(\xi_i)_{i\ge0}$ in the unit ball of ℓ^1. This cannot easily be deduced from (14.26) alone, or even with the help of the theorem of Scheffé.

[3] The theorem says that as long as the sequence $(x_n)_{n\ge1}$ is composed of distributions, and x is another distribution, convergence of coordinates of $(x_n)_{n\ge1}$ to the coordinates of x implies convergence of $(x_n)_{n\ge1}$ to x in the norm of ℓ^1 (see e.g., [11, 12, 15] for more details). To recall, $(\xi_i)_{i\ge0} \in \ell^1$ is said to be a distribution if $\xi_i \ge 0$ and $\sum_{i=0}^\infty \xi_i = 1$.

14.6 Proving Strong Convergence

Suppose that we are given a family $A_\epsilon, \epsilon > 0$ of operators in a Banach space $\mathbb{X}$ and intend to prove that, as $\epsilon \to 0$, the A_ϵ's converge to an operator A. By the Banach–Steinhaus theorem, this convergence is impossible unless the family is composed of equibounded operators (see the reasoning in Section 15.5). Hence, our first step should be to check that the condition of equiboundedness is satisfied. Once we prove that, our task becomes much simpler if, additionally, we know a linearly dense subset of $\mathbb{X}$.

To recall, a subset $\mathbb{X}_0 \subset \mathbb{X}$ is said to be *linearly dense* in $\mathbb{X}$ if the set of linear combinations of its elements is dense in $\mathbb{X}$. This means that, given $x \in \mathbb{X}$ and $\delta > 0$, we can find a linear combination $y = y(x, \delta)$ of elements of $\mathbb{X}_0$ such that $\|x - y\| < \delta$.

14.10 Lemma *Suppose $A_\epsilon, \epsilon > 0$ are equibounded, and $\mathbb{X}_0 \subset \mathbb{X}$ is linearly dense. Then, the A_ϵ's converge strongly to an operator A if and only if for any $x \in \mathbb{X}_0$, $\lim_{\epsilon \to 0} A_\epsilon x = Ax$.*

Proof Necessity of this condition is obvious, and so we focus on sufficiency. By linearity, our assumption implies that $\lim_{\epsilon \to 0} A_\epsilon y = Ay$ for any linear combination of elements of $\mathbb{X}_0$. Moreover, given $x \in \mathbb{X}$ and $\delta > 0$ we can find a linear combination of elements of $\mathbb{X}_0$ such that $\|x - y\| < \delta$, and then

$$\|A_\epsilon x - Ax\| \le \|A_\epsilon x - A_\epsilon y\| + \|A_\epsilon y - Ay\| + \|Ay - Ax\|.$$

Since the operators $A_\epsilon, \epsilon > 0$ are equibounded, there is a constant M such that the first term on the right-hand side does not exceed $M\|x - y\| < M\delta$, and the third term does not exceed $\|A\| \, \|y - x\| < \|A\|\delta$. By taking ϵ small enough we can also make the second term smaller than δ. This shows that the upper limit of the left-hand side above is no greater than $\delta(M + 1 + \|A\|)$. The proof is now complete, δ being arbitrary. □

As an example, think of the space $C[0, \infty]$ (introduced in Exercise 7.5) and of the operators $A_\epsilon, \epsilon > 0$ defined as follows:

$$A_\epsilon x(t) = \epsilon^{-1} \int_t^{t+\epsilon} x(s) \, \mathrm{d}s, \qquad x \in C[0, \infty], t \ge 0. \tag{14.31}$$

We want to prove that the A_ϵ's converge, as $\epsilon \to 0$, to the identity operator in $C[0, \infty]$.

To this end, we estimate first

$$|A_\epsilon x(t)| \le \epsilon^{-1} \int_t^{t+\epsilon} |x(s)| \, \mathrm{d}s \le \epsilon^{-1} \int_t^{t+\epsilon} \|x\| \, \mathrm{d}s = \|x\|, \qquad x \in C[0, \infty], t \ge 0.$$

This shows that $\|A_\epsilon\| \leq 1$, and so these operators are equibounded. (By considering $x(t) := 1$ for all $t \geq 0$ we check in fact that $\|A_\epsilon\| = 1$.) Next, we recall from Exercise 7.5 that the set of functions $e_\lambda, \lambda \geq 0$, defined by $e_\lambda(t) = \mathrm{e}^{-\lambda t}$, $t \geq 0$ is linearly dense in $C[0,\infty]$. Moreover, we have

$$A_\epsilon e_0 = e_0 \qquad \text{and} \qquad A_\epsilon e_\lambda = \frac{1-\mathrm{e}^{-\epsilon\lambda}}{\epsilon\lambda} e_\lambda; \tag{14.32}$$

it thus follows that $\lim_{\epsilon\to 0} A_\epsilon \mathrm{e}_\lambda = \mathrm{e}_\lambda, \lambda \geq 0$, and this, by Lemma 14.10, completes the proof of our claim.

For a variant of this example in $L^1(\mathbb{R}^+)$, we note that $e_\lambda, \lambda > 0$, when seen as members of this space, form a linearly dense set there also. (e_0, of course, does not belong to $L^1(\mathbb{R}^+)$.) Since, using Fubini's theorem, we can estimate as follows,

$$\begin{aligned}
\int_0^\infty \left| \int_t^{t+\epsilon} x(s)\,\mathrm{d}s \right| \mathrm{d}t &\leq \int_0^\infty \int_t^{t+\epsilon} |x(s)|\,\mathrm{d}s\,\mathrm{d}t \\
&= \int_0^\epsilon \int_0^s \mathrm{d}t\,|x(s)|\,\mathrm{d}s + \int_\epsilon^\infty \int_{s-\epsilon}^s \mathrm{d}t\,|x(s)|\,\mathrm{d}s \\
&= \int_0^\epsilon s|x(s)|\,\mathrm{d}s + \epsilon \int_\epsilon^\infty |x(s)|\,\mathrm{d}s \leq \epsilon\|x\|,
\end{aligned}$$

counterparts of the operators A_ϵ of (14.31) in $L^1(\mathbb{R}^+)$ are contractions too. It follows that these counterparts converge, as $\epsilon \to 0$, to the identity operator in $L^1(\mathbb{R}^+)$, because the second relation in (14.32) remains in force.

14.7 Exercises

Exercise 14.1. Find your own example showing that a sequence $(x_n)_{n\geq 1}$ may diverge even if $\left(\frac{x_1+\cdots+x_n}{n}\right)_{n\geq 1}$ converges.

Exercise 14.2. Use (12.6) to show that the operators $A(t)$ of Section 14.1 do not converge to the identity I in the operator norm, as $t \to 0$.

Exercise 14.3. Suppose $A_n, n \geq 1$ are contractions. Show that they cannot converge strongly to an operator A with $\|A\| > 1$. **Hint:** Think of an x of norm one such that $\|Ax\| > \frac{1}{2}(\|A\| + 1) > 1$.

Exercise 14.4. Show by example that a sequence of operators of norm 1 can converge strongly to the zero operator. **Hint:** In ℓ^1 let $A_n, n \geq 1$ replace the first n coordinates of a sequence by n zeros.

Exercise 14.5. The space c_0 is a completion of c_{00}. Show that the functionals A_n of (14.4), even though they are well defined in c_0, do not converge strongly in this space.

Exercise 14.6. Let $C[0,\infty]$ be the space of continuous functions $x\colon [0,\infty) \to \mathbb{R}$ that have finite limits at infinity, and let $A_n, n \geq 1$ be the operators in this space defined by

$$A_n x(t) = x\left(t + 1/n\right), \qquad t \geq 0.$$

Prove that the sequence $(A_n)_{n\geq 1}$ converges to the identity operator strongly, but not in the operator norm.

Exercise 14.7. Deduce (14.14)–(14.16) from Euler's identities (7.4).

Exercise 14.8. Prove that for any $t \in \mathbb{R}$ and 2π-periodic continuous function $y\colon \mathbb{R} \to \mathbb{R}$, equation (14.21) holds true. **Hint:** It suffices to prove this for $t \in (-\pi,\pi)$. If, for instance, $t > 0$, $\int_{t-\pi}^{t+\pi} y(s)\,\mathrm{d}s = \int_{t-\pi}^{\pi} y(s)\,\mathrm{d}s + \int_{\pi}^{t+\pi} y(s)\,\mathrm{d}s$ and $\int_{\pi}^{t+\pi} y(s)\,\mathrm{d}s = \int_{-\pi}^{t-\pi} y(s)\,\mathrm{d}s$.

Exercise 14.9. Prove (14.30) by induction.

Exercise 14.10. Consider the operators $A_\epsilon, \epsilon > 0$, defined in $C[0,\infty]$ by

$$A_\epsilon x(t) = \frac{2}{\epsilon^2}\int_t^{t+\epsilon}\int_t^s x(u)\,\mathrm{d}u\,\mathrm{d}s = \frac{2}{\epsilon^2}\int_t^{t+\epsilon}(t+\epsilon-s)x(s)\,\mathrm{d}s,\ x \in C[0,\infty], t \geq 0,$$

and prove their convergence to the identity operator. Prove also a variant of this exercise in $L^1(\mathbb{R}^+)$. Do not forget to check that the operators defined above, and those defined in (14.31), do map $C[0,\infty]$ into itself.

Exercise 14.11. ▲ Prove that the operators $A_\epsilon, \epsilon > 0$ in $L^1(\mathbb{R})$ given by

$$A_\epsilon x(t) = \epsilon^{-1}\int_t^{t+\epsilon} x(s)\,\mathrm{d}s, \qquad x \in L^1(\mathbb{R}), t \geq 0,$$

are contractions that converge, as $\epsilon \to 0$, to the identity operator in this space. **Hint:** Argue first that it suffices to check that

$$\lim_{\epsilon\to 0} A_\epsilon e_\lambda^{\text{even}} = e_\lambda^{\text{even}} \qquad \text{and} \qquad \lim_{\epsilon\to 0} A_\epsilon e_\lambda^{\text{odd}} = e_\lambda^{\text{odd}}, \lambda > 0, \tag{14.33}$$

where $e_\lambda^{\text{even}}(t) = \mathrm{e}^{-\lambda|t|}, t \in \mathbb{R}$ whereas $e_\lambda^{\text{odd}}(t) = \mathrm{e}^{-\lambda t}$ for $t \geq 0$ and $e_\lambda^{\text{odd}}(t) = \mathrm{e}^{\lambda t}$ for $t < 0$. To prove the first part of (14.33), check to see that

$$A_\epsilon e_\lambda^{\text{even}}(t) = \begin{cases} \frac{1-\mathrm{e}^{-\lambda\epsilon}}{\epsilon\lambda} e_\lambda^{\text{even}}(t), & t \geq 0, \\ \frac{\mathrm{e}^{\lambda\epsilon}-1}{\epsilon\lambda} e_\lambda^{\text{even}}(t), & t \leq -\epsilon, \end{cases}$$

and that $\lim_{\epsilon\to 0}\int_{-\epsilon}^{0} A_\epsilon e_\lambda^{\text{even}}(t)\,\mathrm{d}t = 0$, because $0 \leq A_\epsilon e_\lambda^{\text{even}}(t) \leq 1$.

☞ CHAPTER SUMMARY

Although there is a particular beauty in the statement that the space of bounded linear operators in a Banach space is a Banach space itself, the norm in this space is more often than not too strong to encompass the more delicate convergence theorems of contemporary mathematics. Strong convergence is a notion that is more suitable for such purposes. We exemplify this by studying two classical theorems: Bernstein's approximation of continuous functions by polynomials and the theorem of Fejér on convergence of Fourier series. In both cases the operators involved converge strongly but not in the operator norm. Before doing that, however, we discuss the theorem of Banach and Steinhaus, which says that for a sequence $(A_n)_{n\ge 1}$ of bounded linear operators defined in a Banach space $\mathbb{X}$, condition $\sup_{n\ge 1} \|A_n x\| < \infty$ for all $x \in \mathbb{X}$, implies $\sup_{n\ge 1} \|A_n\| < \infty$. This result ensures in particular that in Banach spaces strong convergence of bounded linear operators implies boundedness of the limit operator. The chapter also covers the famous Poisson approximation to the binomial, the only example of a limit theorem of probability known to the author that can in fact be stated in the framework of norm convergence of operators.

15

We Go Deeper, Deeper We Go (into the Structure of Complete Spaces)

There are reasons why *true* functional analysts should not and would not be satisfied with Sokal's elementary proof of the Banach–Steinhaus theorem. Some of them would be in fact inclined to say that this proof is "lock-picking": it is the Baire category theorem that is the real key. They would maintain that the Baire category theorem provides an intrinsic description of a complete metric space, and, hence, all important functional analytic results are in essence hidden in this description and should be derived from this description.

This or similar rationale led S. Saks, who was the referee of the original paper of Banach and Steinhaus [7], to suggest an alternative proof[1] – that is, the proof based on Baire's theorem that can nowadays be found in almost all books on functional analysis. There are reasons to believe that their first proof used the gliding hump argument, which was quite popular at that time, and well known to Banach and Steinhaus. But we will probably never know this for sure, since the manuscript was destroyed during World War II. (See [13] pp. 52–53 for more on this subject.)

Even though I am attracted by the elegance and simplicity of Sokal's proof, the arguments presented above are weighty. Therefore, in this chapter, we indeed go deeper into the topological structure of Banach spaces by discussing briefly the celebrated category theorem of Baire. From this theorem we then deduce (again) the Banach–Steinhaus uniform boundedness theorem, and then cannot refrain from proving another of its profound consequences: the open mapping theorem, one of the cornerstones of functional analysis and operator theory. A proof of the uniform boundedness principle that is based on the gliding hump argument can be found, for example, in [13] or [31]. Another fundamental theorem of functional analysis and operator theory, the closed graph theorem, is discussed in Section 15.9.

[1] What a quality report that was!

15.1 Baire's Category Theorem

Let's start from scratch. A subset, say S, of a metric space $\mathbb{X}$ is termed nowhere dense if and only if its closure does not contain any open ball. The Baire category theorem says that *a complete space can not be represented as a countable union of nowhere dense sets*. (Of course, in this statement, we tacitly assume that the space we consider is non-empty.) The name of the theorem comes from the fact that sets that can be represented as a countable union of nowhere dense sets (e.g., countable sets) are termed sets of the first category. In this terminology, Baire's theorem states that complete spaces are not of the first category; they are sets of the second category.

We start the proof of the theorem by noting that if S is nowhere dense, then any open ball B contains a ball B' such that $B' \cap S$ is empty. Indeed, if we suppose that any ball B' that is a subset of B contains an element of S, then it is easy to see that every point of B belongs to the closure of S. To this end it suffices to consider, for every point x of B, the sequence of balls $B(x, \frac{1}{n})$ with n large enough to have $B(x, \frac{1}{n}) \subset B$. But the entire B cannot be contained in the closure of S, since S is nowhere dense. This contradiction shows that at least one open ball contained in B has no common points with S.

With this preparation under our belt, assume a complete metric space $\mathbb{X}$ is a union of nowhere dense sets $S_n, n \geq 1$:

$$\mathbb{X} = \bigcup_{n \in \mathbb{N}} S_n. \tag{15.1}$$

Let B_1 be the open ball with radius 1 and center 0. Since S_1 is nowhere dense, B_1 contains, as we have just observed, an open ball B_2 that is disjoint with S_1. We can actually assume that the radius of B_2 is smaller than $\frac{1}{2}$, and that the closure of B_2 is contained in B_1 and disjoint with S_1; it is just a matter of taking a smaller radius, if necessary. This procedure can be repeated: we can find an open ball B_3 of radius smaller than $\frac{1}{3}$ such that its closure is contained in B_2 and disjoint with S_2. More generally, having found an open ball B_n, we can find an open ball B_{n+1} with radius smaller than $\frac{1}{n+1}$, whose closure is contained in B_n and at the same time disjoint with S_n.

This, however, leads us straight to a contradiction. For, the centers, say x_n, of balls B_n are seen to form a Cauchy sequence: the distance between x_n and x_m does not exceed $\frac{1}{n}$ whenever $m \geq n$. Since the balls form a decreasing sequence in that $B_{n+1} \subset B_n$ for $n \geq 1$, the limit of $(x_n)_{n \geq 1}$ belongs to the closure of B_n for each $n \geq 2$, and thus is not a member of S_{n-1}. This, of course, contradicts (15.1).

Figure 15.1 A set that is not meagre.

15.2 Differentiable Functions Form a *Meagre* Set in $C[a,b]$

From the perspective of Baire's theorem, sets of the first category, known also as *meagre* sets, are in a sense small (see Figure 15.1). Since it is hard for us to *imagine*, not to mention *draw*, the graph of a function that is nowhere differentiable, it is often found surprising[2] that the set of functions that do have a derivative somewhere in $[0,1]$ is meagre in $C[0,1]$. Here is, however, a proof of this curious fact.[3]

Consider first the sets $D_k \subset C[0,1]$ for $k = 2,3,\ldots$ defined as follows. An $x \in C[0,1]$ belongs to D_k if there is a $t \in [0, 1-\frac{1}{k}]$ such that for all $h \in (0, 1-t]$,

$$|x(t+h) - x(t)| \le kh. \tag{15.2}$$

[2] Readers who fancy category theory may want to consult [26] for a proof that the set of stochastic semigroups that are not asymptotically stable is meagre in the space of all stochastic semigroups. Needless to say, proving that a given stochastic semigroup is asymptotically stable is a task in itself.

[3] S. Mazurkiewicz writes [29] that the question of category of the set of differentiable functions as a subset of the space of continuous functions was posed in 1929 by H. Steinhaus (see [39, p. 81]). Two answers were almost simultaneously given by S. Mazurkiewicz [29] and S. Banach [6] and articles of these two authors appeared in the same, third, issue of Studia Mathematica in 1931. (I thank Krzysztof Ciesielski for these references.) We closely follow Banach's argument because is seems to be considerably simpler.

We will show that:

1. Each D_k is closed.
2. Each D_k is nowhere dense.
3. The set $\mathfrak{D}$ of functions $x \in C[0,1]$ that have a finite right-hand derivative at at least one point of $[0,1)$ is contained in $D := \bigcup_{k\geq 2} D_k$.

Points 1. and 2. combined show that D is *meagre*, and then 3. shows that so is $\mathfrak{D}$.

To prove 1., we fix a k and consider a sequence $(x_n)_{n\geq 1}$ of elements of D_k converging to an $x \in C[0,1]$. By definition there are $t_n \in [0, 1-\frac{1}{k}]$ such that

$$|x_n(t_n + h) - x_n(t_n)| \leq kh \tag{15.3}$$

whenever $t_n + h \leq 1$ and $h > 0$. Since $[0, 1-\frac{1}{k}]$ is compact, $(t_n)_{n\geq 1}$ has a convergent subsequence. Without loss of generality and to simplify notations, we assume that the entire sequence converges to, say, a $t \in [0, 1-\frac{1}{k}]$.

Next, let $h > 0$ be such that $t + h \leq 1$. Then $h_n := h + t - t_n$ is positive for sufficiently large n, and $t_n + h_n = t + h \leq 1$. Hence, by (15.3),

$$|x_n(t_n + h_n) - x_n(t_n)| \leq kh_n. \tag{15.4}$$

We claim now that, as $n \to \infty$, the left-hand side converges to $|x(t+h) - x(t)|$. To see this, given $\epsilon > 0$ we find n_0 such that $\|x - x_n\| < \frac{\epsilon}{3}$ for $n \geq n_0$. For such n's we can write

$$\begin{aligned}|x_n(t_n + h_n) - &x_n(t_n) - x(t+h) + x(t)| \\ &\leq |x_n(t_n + h_n) - x(t_n + h_n)| + |x(t_n + h_n) - x(t+h)| \\ &\quad + |x_n(t_n) - x(t_n)| + |x(t_n) - x(t)| \\ &\leq \tfrac{2}{3}\epsilon + |x(t_n + h_n) - x(t+h)| + |x(t_n) - x(t)|,\end{aligned}$$

and, by continuity of x, we can make the last two terms above smaller than $\frac{1}{3}\epsilon$ by taking n perhaps even larger. This shows that $\lim_{n\to\infty}[x_n(t_n + h_n) - x_n(t_n)] = x(t+h) - x(t)$ and so proves our claim.

Thus, letting n tend to ∞ in (15.4) yields

$$|x(t+h) - x(t)| \leq kh,$$

proving that $x \in D_k$, and thus showing that D_k is closed.

Turning to 2., we assume toward a contradiction that a certain D_k contains an open ball, say, B. Then, by the Weierstrass theorem there is a polynomial $p \in B$ and then a radius $r > 0$ such that for any $y \in C[0,1]$, $\|y - p\| < r$ implies $y \in B \subset D_k$. But, as we shall now see, this cannot be true.

Namely, see Figure 15.2, one may construct a $z \in C[0,1]$ such that $\|z\| < r$, whereas the right-hand derivatives of z at all points $t \in [0,1)$ exist, are finite, but are larger than $\|p'\| + 2k$, where p' is the derivative of p. Then for $y :=$

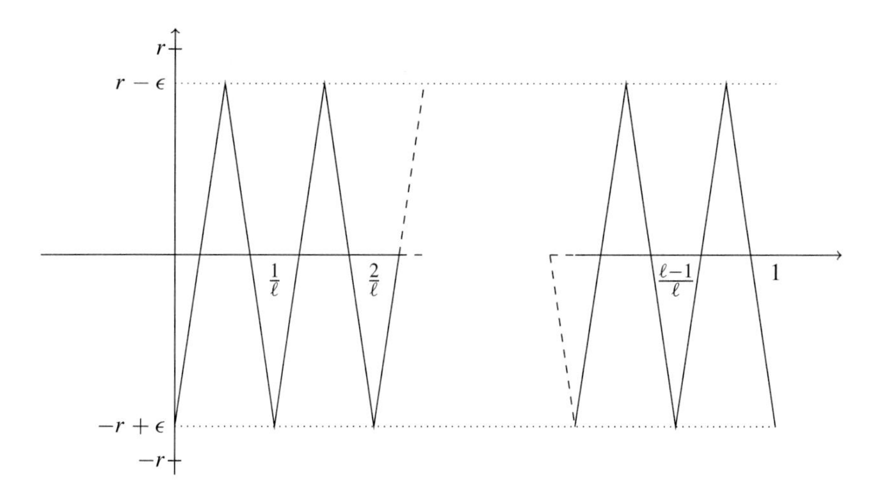

Figure 15.2 This function has right-sided derivatives at all $t \in [0, 1)$ having absolute values equal to $4\ell(r-\epsilon)$, and the integer ℓ may be chosen arbitrarily large.

$p + z$, we have $y'_+(t) = p'(t) + z'_+(t), t \in [0, 1)$ (z'_+ and y'_+ are right-hand derivatives), and thus

$$|y'_+(t)| \geq |z_+(t)| - |p'(t)| \geq |z_+(t)| - \|p'\| \geq 2k, \qquad t \in [0, 1).$$

It follows that for each t there is an $h > 0$ such that $t + h \leq 1$ and $\frac{|y(t+h)-y(t)|}{h} > k$, excluding the possibility of z belonging to D_k. Since this is despite the fact that $\|y - p\| = \|z\| < r$, we have the required contradiction.

We are thus left with showing that $\mathfrak{D} \subset D$. To this end, think of an x that has a finite right-hand derivative at a $t \in [0, 1)$. Then, the continuous function $(0, 1-t] \ni h \mapsto \frac{x(t+h)-x(t)}{h}$ can be extended to a continuous function on $[0, 1-t]$. Therefore, there is an $M > 0$ such that

$$\frac{|x(t+h) - x(t)|}{h} \leq M$$

for $h \in (0, 1-t]$. Thus, for k so large that $t \leq 1 - \frac{1}{k}$ and $k \geq M$, x belongs to D_k.

Hence, the position of $\mathfrak{D}$ in $C[0, 1]$ is comparable to the position of the set of rational numbers in $\mathbb{R}$. Both are dense, and yet meagre. Isn't it surprising?

15.3 The Banach–Steinhaus Theorem II

Since a Banach space is a particular case of a complete metric space, we can apply Baire's category theorem to prove a number of results pertaining to Banach spaces. In this section, we show how to use it to prove the Banach–Steinhaus theorem, which we present now in a slightly different form.

15.1 Theorem (Banach–Steinhaus Theorem) *Let $A_n, n \geq 1$ be bounded linear operators from a normed space $\mathbb{X}$ to a normed space $\mathbb{Y}$. Also, let $\mathbb{X}_0$ be the subspace of $x \in \mathbb{X}$ such that*

$$\sup_{n \geq 1} \|A_n x\| < \infty.$$

If $\mathbb{X}_0$ is of the second category, then $\mathbb{X}_0 = \mathbb{X}$ and $\sup_{n\geq 1} \|A_n\| < \infty$.

Proof Let $S_n := \{x \in \mathbb{X}; \sup_{k\geq 1} \|A_k x\| \leq n\} = \bigcap_{k\geq 1}\{x \in \mathbb{X}; \|A_k x\| \leq n\}$. Then, each S_n is closed (the operators $A_k, k \geq 1$ being bounded) and $S_n \subset \mathbb{X}_0$. Furthermore, for $x \in \mathbb{X}_0$ there is an n such that $x \in S_n$. This means that $\mathbb{X}_0 = \bigcup_{n\geq 1} S_n$.

Since $\mathbb{X}_0$ is by assumption of the second category, at least one of S_n is not nowhere dense. In other words, for a certain $\ell \in \mathbb{N}$, there exists an open ball B contained in S_ℓ. Let x be the center of this ball and let $r > 0$ be its radius. Consider a non-zero $y \in \mathbb{X}$ and the related vector $z := x + \frac{r}{2\|y\|} y \in B$. We have

$$\|A_n y\| = \frac{2\|y\|}{r} \|A_n z - A_n x\| \leq \frac{2\|y\|}{r} (\|A_n z\| + \|A_n x\|) \leq \frac{4\ell}{r}\|y\|, \quad n \geq 1, \tag{15.5}$$

because both z and x belong to $B \subset S_\ell$. It follows that y belongs to $\mathbb{X}_0$. Since y is arbitrary, this proves the first part of the thesis. But (15.5) shows more: it shows that $\sup_{n\in\mathbb{N}} \|A_n\| \leq \frac{4\ell}{r}$. This completes the proof. □

15.2 Remark

(a) If $\mathbb{X}$ is complete and $\mathbb{X}_0 = \mathbb{X}$, $\mathbb{X}_0$ is automatically of the second category, by the Baire category theorem. Hence, this form of the Banach–Steinhaus theorem includes the previous one.

(b) It is an immediate corollary of this new version that if $\sup_{n\geq 1} \|A_n\| = \infty$, the set $\mathbb{X}_0$ must be 'small,' that is, not of the second category. In particular, $\mathbb{X} \setminus \mathbb{X}_0$ must be non-empty.

15.4 Can the Fourier Series Converge in $C_p[-\pi, \pi]$?

In Section 14.4 we have proved, following Fejér, that for any $x \in C_p[-\pi,\pi]$, the functions

$$\sigma_n := \frac{s_0 + \cdots + s_{n-1}}{n}, \qquad n \geq 1,$$

where the partial sums $s_0, s_1, s_2, \ldots$ are defined by (14.11)–(14.12), converge to x as $n \to \infty$. We are now ready to answer the question of whether we could have proved a stronger result: that the partial sums $s_n, n \geq 0$ themselves converge to x. The answer is in the negative: by the Banach–Steinhaus theorem this convergence cannot hold on the entire $C_p[-\pi,\pi]$. We cannot even expect the numerical sequence $(s_n(0))_{n\geq 0}$ to converge for all x! This is to say that there are x's for which $\lim_{n\to\infty} s_n(0)$ does not exist. In fact, such x's form a rather large set, for the set of x for which $(s_n(0))_{n\geq 0}$ is bounded is not of the second category.

To see this, consider the functionals $F_n \colon C_p[-\pi,\pi] \to \mathbb{R}$ given by (see (14.19))

$$x \mapsto F_n x := s_n(0) = \frac{1}{\pi} \int_{-\pi}^{\pi} x(s) k_n(s) \, \mathrm{d}s,$$

where, by (14.17), $k_n(s) = \frac{\sin(n+\frac{1}{2})s}{2\sin\frac{s}{2}}$. We will show that

$$\sup_{n\geq 1} \|F_n\| = \infty. \tag{15.6}$$

By Remark 15.2 (b) this indeed implies that the set of x for which $(s_n(0))_{n\geq 0}$ is bounded is not of the second category. In particular, this excludes the possibility that $\lim_{n\to\infty} s_n(0)$ exists and is finite for all $x \in C_p[-\pi,\pi]$.

Our first step is to note that

$$\|F_n\| = \frac{1}{\pi} \int_{-\pi}^{\pi} |k_n(t)| \, \mathrm{d}t = \frac{2}{\pi} \int_{0}^{\pi} |k_n(t)| \, \mathrm{d}t. \tag{15.7}$$

We proved such a relation at the end of Section 12.2.3, when we examined the functional from Example 12.11. Of course, $C_p[-\pi,\pi]$ is not quite the same as $C[-\pi,\pi]$, but the argument presented there works for $C_p[-\pi,\pi]$ also, because k_n itself is a member of $C_p[-\pi,\pi]$.

Having established (15.7) (its second part is a direct consequence of the fact that k_n is even), think of the intervals

$$I_\ell := \left(\frac{(\ell-1)\pi}{n+\frac{1}{2}}, \frac{\ell\pi}{n+\frac{1}{2}} \right], \qquad \ell = 1, \ldots, n$$

contained in $[0,\pi]$. For $t \in I_\ell$, $\sin\frac{t}{2} \le \frac{t}{2} \le \frac{\ell\pi}{2n+1}$. Hence,

$$\int_{I_\ell} |k_n(t)|\,\mathrm{d}t = \int_{\frac{(\ell-1)\pi}{n+\frac{1}{2}}}^{\frac{\ell\pi}{n+\frac{1}{2}}} |k_n(t)|\,\mathrm{d}t \ge \frac{2n+1}{\ell\pi}\int_{\frac{(\ell-1)\pi}{n+\frac{1}{2}}}^{\frac{\ell\pi}{n+\frac{1}{2}}} |\sin(n+\tfrac{1}{2})t|\,\mathrm{d}t$$

$$= \frac{2}{\ell\pi}\int_{(\ell-1)\pi}^{\ell\pi} |\sin t|\,\mathrm{d}t = \frac{2}{\ell\pi}\int_0^{\pi} \sin t\,\mathrm{d}t = \frac{4}{\ell\pi}.$$

Since $\bigcup_{\ell=1}^n I_\ell \subset [0,\pi]$ and I_ℓ are pairwise disjoint, we obtain

$$\|F_n\| \ge \frac{4}{\pi^2}\sum_{\ell=1}^{n}\frac{1}{\ell},$$

and (15.6) follows.

We have proved that there is a relatively large set of $x \in C_{\mathrm{p}}[-\pi,\pi]$ for which $\lim_{n\to\infty} s_n(0)$ does not exist: the sequence $(s_n(0))_{n\ge 0}$ is not even bounded. This is despite the fact that we did not construct any such x explicitly.

15.5 Exponential Growth of Semigroups of Operators

Here is another typical argument based on the Banach–Steinhaus theorem, which also shows how to apply the theorem to uncountable families of operators.

A family $\{T(t), t \ge 0\}$ of bounded linear operators in a Banach space $\mathbb{X}$ is said to be a semigroup if $T(0)$ is the identity operator in $\mathbb{X}$ and

$$T(t)T(s) = T(t+s), \qquad s,t \ge 0. \tag{15.8}$$

It is said to be a C_0-semigroup (or: a *semigroup of class* C_0 or a *strongly continuous semigroup*) if, additionally, $\lim_{t\to 0} T(t)x = x$, that is, $\lim_{t\to 0}\|T(t)x - x\| = 0$, for all $x \in \mathbb{X}$. (This means that $T(t)$ converges to the identity operator in the strong topology.)

As it turns out, for each semigroup of class C_0 there are $M \ge 1$ and $\omega \in \mathbb{R}$ such that

$$\|T(t)\| \le M\mathrm{e}^{\omega t}, \qquad t \ge 0, \tag{15.9}$$

and the proof employs the Banach–Steinhaus theorem in a crucial way, as is seen below.

More specifically, the proof hinges on the observation, which is a consequence of the Banach–Steinhaus theorem, that there is a $\delta > 0$ such that the supremum

$$M := \sup_{t\in[0,\delta]} \|T(t)\|$$

is finite. Indeed, if this is not the case, then for any $n \geq 1$ there is a $t_n > 0$ such that $t_n < \frac{1}{n}$ whereas $\|T(t_n)\| \geq n$. Since $\lim_{n\to\infty} t_n = 0$, this, by the Banach–Steinhaus theorem, contradicts the assumption that $\lim_{t\to 0} T(t)x = x$. To phrase it differently: our assumption implies $\lim_{n\to\infty} T(t_n)x = x$, and a fortiori $\sup_{n\geq 1} \|T(t_n)x\| < \infty$ for all $x \in \mathbb{X}$. The Banach–Steinhaus theorem then says that $\sup_{n\geq 1} \|T(t_n)\| < \infty$, contradicting $\|T(t_n)\| \geq n$.

The rest is simple: given $t \geq 0$ we find the $n \in \mathbb{N}\cup\{0\}$ such that $t = n\delta + t'$ for a $t' \in [0,\delta)$ and obtain:

$$\begin{aligned}\|T(t)\| &= \|[T(\delta)]^n T(t')\| \leq \|T(\delta)\|^n \|T(t')\| \leq M^n M = M\mathrm{e}^{n\ln M}\\ &\leq M\mathrm{e}^{(n\delta+t')\frac{\ln M}{\delta}} = M\mathrm{e}^{\frac{\ln M}{\delta}t}.\end{aligned}$$

This is our claim with $\omega := \frac{\ln M}{\delta}$.

The inequality (15.9), in turn, allows us to prove that the map $[0,\infty) \ni t \mapsto T(t)$ is continuous in the strong topology. Indeed, continuity at $t = 0$ is guaranteed by the definition of the C_0-semigroup, and the right-hand continuity at a $t > 0$ is a direct consequence of

$$\|T(t+h)x - T(t)x\| = \|T(t)[T(h)x - x]\| \leq \|T(t)\|\,\|T(h)x - x\|, \qquad h \geq 0, x \in \mathbb{X},$$

because $\|T(t)\| < \infty$ and $\lim_{h\to 0} \|T(h)x - x\| = 0$ by assumption for any $x \in \mathbb{X}$. It is when we wish to prove the left-hand continuity that we need to use (15.9), and we do this as follows. For $t > 0$ and $h \in (0,t)$,

$$\begin{aligned}\|T(t-h)x - T(t)x\| &= \|T(t-h)[x - T(h)x]\| \leq \|T(t-h)\|\,\|x - T(h)x\|\\ &\leq M\mathrm{e}^{\omega(t-h)}\|x - T(h)x\|.\end{aligned}$$

The right-hand side converges to 0, as $h \to 0$, because $M\mathrm{e}^{\omega(t-h)}$ does not exceed M when $\omega \leq 0$ and does not exceed $M\mathrm{e}^{\omega t}$ when $\omega > 0$.

15.6 The Open Mapping Theorem

A function $f\colon X \to Y$ between two topological spaces X and Y is said to be open if it maps open sets to open sets, that is, if for every open set $U \subset X$, its image $f(U)$ is open in Y. This definition is similar to that of a continuous function, but in fact describes a rather different class of maps. For example, if a map f is one-to-one and onto, openness of f is equivalent to continuity of its inverse f^{-1}, but apparently says nothing about the continuity of f itself.

The open mapping theorem states that a bounded linear map $A\colon \mathbb{X} \to \mathbb{Y}$, where $\mathbb{X}$ and $\mathbb{Y}$ are Banach spaces, is necessarily open as long as A is onto.

15.3 Theorem *Suppose $\mathbb{X}$ and $\mathbb{Y}$ are Banach spaces and a bounded linear map $A\colon \mathbb{X} \to \mathbb{Y}$ is onto. Then, the image of the open unit ball in $\mathbb{X}$ with center at $0 \in \mathbb{X}$ contains an open ball in $\mathbb{Y}$ with center at $0 \in \mathbb{Y}$. That is, there is a $\rho > 0$ such that*

$$B_{\mathbb{Y}}(0,\rho) \subset A\left(B_{\mathbb{X}}(0,1)\right), \tag{15.10}$$

where, by definition, for any $r > 0$,

$$B_{\mathbb{Y}}(0,r) = \{y \in \mathbb{Y}; \|y\| < r\} \qquad \textit{and} \qquad B_{\mathbb{X}}(0,r) = \{x \in \mathbb{X}; \|x\| < r\}.$$

15.4 Remark

(i) The statement of the theorem implies that A is open. To see this, consider an open subset U of $\mathbb{X}$ and let y belong to $A(U)$. We need to check that there is a ball with center at y that is contained in $A(U)$, that is, there is an $\epsilon > 0$ such that $B_{\mathbb{Y}}(y,\epsilon) \subset A(U)$. Let $x \in U$ be such that $Ax = y$. Since U is open, there is an ε such that $B_{\mathbb{X}}(x,\varepsilon) \subset U$. Then, by (15.10),

$$B_{\mathbb{Y}}(0,\varepsilon\rho) \subset A(B_{\mathbb{X}}(0,\varepsilon)).$$

It follows that[4]

$$B_{\mathbb{Y}}(y,\varepsilon\rho) = y + B_{\mathbb{Y}}(0,\varepsilon\rho) \subset Ax + A(B_{\mathbb{X}}(0,\varepsilon)) \subset A(B_{\mathbb{X}}(x,\varepsilon)) \subset A(U),$$

that is, that $B_{\mathbb{Y}}(y,\epsilon) \subset A(U)$ for $\epsilon := \varepsilon\rho$.

(ii) As a by-product of the proof presented below, the theorem remains valid if we assume merely that the image of A is of the second category in $\mathbb{Y}$.

Proof of Theorem 15.3 This classic proof naturally splits into two parts.

(a) Since A is onto, and $\mathbb{X} = \bigcup_{n\ge 1} B_{\mathbb{X}}(0,n)$, we have

$$\mathbb{Y} = \bigcup_{n\ge 1} A(B_{\mathbb{X}}(0,n)).$$

Therefore, by the Baire category theorem, at least one of the sets $A(B_{\mathbb{X}}(0,n)), n \ge 1$ is not nowhere dense, $\mathbb{Y}$ as a Banach space being not meagre. Let $n_0 \ge 1$, $y \in \mathbb{Y}$ and $\epsilon > 0$ be such that

$$B_{\mathbb{Y}}(y,\epsilon) \subset cl\, A(B_{\mathbb{X}}(0,n_0)).$$

The set on the right-hand side here is symmetric: a $\widetilde{y}$ belongs to this set whenever $-\widetilde{y}$ does. It follows that

$$B_{\mathbb{Y}}(-y,\epsilon) \subset cl\, A(B_{\mathbb{X}}(0,n_0)),$$

[4] To recall, for a set $S \subset \mathbb{Y}$ and a $y \in \mathbb{Y}$, $y + S$ is the set of vectors of the form $y + \widetilde{y}$ where $\widetilde{y}$ belongs to S.

too. The set on the right is also convex, that is, for any two elements, say y_1, y_2, of this set, and for any $\alpha \in [0, 1]$, the convex combination $\alpha y_1 + (1 - \alpha)y_2$ belongs to this set. Hence, combining these two inclusions with the fact that any $z \in \mathbb{Y}$ of norm smaller than ϵ can be written as

$$z = \frac{1}{2}(y + z) + \frac{1}{2}(-y + z),$$

where $y + z \in B_{\mathbb{Y}}(y, \epsilon)$ and $-y + z \in B_{\mathbb{Y}}(-y, \epsilon)$, we see that

$$B_{\mathbb{Y}}(0, \epsilon) \subset cl\ A(B_{\mathbb{X}}(0, n_0)).$$

This, however, means that for $\rho := \frac{\epsilon}{2n_0}$,

$$B_{\mathbb{Y}}(0, 2\rho) \subset cl\ A(B_{\mathbb{X}}(0, 1)). \tag{15.11}$$

(b) Relation (15.11), when combined with completeness of $\mathbb{X}$, is a key to the proof of (15.10). Here are the details.

It is useful to note that (15.11) implies that for all $n \geq 0$,

$$B_{\mathbb{Y}}(0, \tfrac{\rho}{2^{n-1}}) \subset cl\ A(B_{\mathbb{X}}(0, \tfrac{1}{2^n})). \tag{15.12}$$

Let y belong to $B_{\mathbb{Y}}(0, \rho)$. By (15.12) with $n = 1$, this y can be approximated to any given accuracy by elements of $A(B_{\mathbb{X}}(0, \frac{1}{2}))$. In particular, there is an $x_1 \in \mathbb{X}$ of norm smaller than $\frac{1}{2}$ such that

$$\|y - Ax_1\| < \frac{1}{2}\rho.$$

Thinking next of $y - Ax_1$ instead of y and using (15.12) with $n = 2$, we see that there is an x_2 of norm smaller than $\frac{1}{4}$ such that

$$\|y - Ax_1 - Ax_2\| < \frac{1}{4}\rho.$$

Repeating this procedure, we see that there is a sequence $(x_n)_{n\geq 1}$ of elements of $\mathbb{X}$ such that

$$\|x_n\| < \frac{1}{2^n} \qquad \text{whereas} \qquad \|y - \sum_{i=1}^{n} Ax_i\| < \frac{\rho}{2^n} \tag{15.13}$$

for all $n \geq 1$. Since, for $k < \ell$,

$$\left\| \sum_{n=1}^{\ell} x_n - \sum_{n=1}^{k} x_n \right\| = \left\| \sum_{n=k+1}^{\ell} x_n \right\| \leq \sum_{n=k+1}^{\ell} \frac{1}{2^n}$$

and the series $\sum_{n=1}^{\infty} \frac{1}{2^n}$ converges, it is clear that $(\widetilde{x}_n)_{n\geq 1}$, where $\widetilde{x}_n := \sum_{i=1}^{n} x_i$, is a Cauchy sequence in $\mathbb{X}$. Thus, $\mathbb{X}$ being complete, $(\widetilde{x}_n)_{n\geq 1}$ has a limit, say x:

$$x := \sum_{n=1}^{\infty} x_n.$$

Letting $n \to \infty$ in the second relation in (15.13) and using the continuity of A we obtain $y = Ax$. Moreover,

$$\|x\| \leq \lim_{n\to\infty} \left(\|x_1\| + \sum_{i=2}^{n} \|x_i\| \right) < \frac{1}{2} + \lim_{n\to\infty} \sum_{i=2}^{n} \frac{1}{2^i} < 1,$$

showing that x belongs to $B_{\mathbb{X}}(0,1)$. Since y is arbitrary, this proves (15.10). □

15.5 Corollary *Suppose $\mathbb{X}$ and $\mathbb{Y}$ are Banach spaces, and that a bounded operator $A\colon \mathbb{X} \to \mathbb{Y}$ is onto and one-to-one, that is, for any $y \in \mathbb{Y}$ there is precisely one $x \in \mathbb{X}$ such that $Ax = y$. Then the inverse operator $A^{-1}\colon \mathbb{Y} \to \mathbb{X}$ is bounded with*

$$\|A^{-1}y\| \leq \frac{1}{\rho}\|y\|, \qquad y \in \mathbb{Y},$$

where ρ is the constant from (15.10).

Proof It suffices to show the thesis for $y \neq 0$. For such a y and $r \in (0,1)$, consider $y_r := \frac{r\rho}{\|y\|} y$. By (15.10), since $\|y_r\| = r\rho < \rho$, there is an $x \in B_{\mathbb{X}}(0,1)$ such that $Ax = y_r$. It follows that $\|A^{-1}y_r\| = \|x\| < 1$, that is, $\|A^{-1}y\| < \frac{1}{r\rho}\|y\|$. Letting $r \to 1$ completes the proof. □

15.7 Operators Bounded Below

Let $\mathbb{X}$ and $\mathbb{Y}$ be normed spaces with norms $\|\cdot\|_{\mathbb{X}}$ and $\|\cdot\|_{\mathbb{Y}}$, respectively. A linear operator $A\colon \mathbb{X} \to \mathbb{Y}$ is said to be *bounded below* if there is a $\delta > 0$ such that

$$\|Ax\|_{\mathbb{Y}} \geq \delta\|x\|_{\mathbb{X}} \qquad \text{for all } x \in \mathbb{X}. \tag{15.14}$$

It is clear that such operators have trivial kernels and thus are injective: $Ax = 0$ implies $x = 0$.

For example, any operator A that has a bounded left inverse is bounded below. To see this, let B be this inverse so that $BAx = x, x \in \mathbb{X}$. Then $\|x\|_{\mathbb{X}} = \|BAx\|_{\mathbb{X}} \leq \|B\|_{\mathcal{L}(\mathbb{Y},\mathbb{X})}\|Ax\|_{\mathbb{Y}}$ for $x \in \mathbb{X}$. This implies (15.14) with $\delta = (\|B\|_{\mathcal{L}(\mathbb{Y},\mathbb{X})})^{-1}$.[5]

[5] $\|B\|_{\mathcal{L}(\mathbb{Y},\mathbb{X})}$ cannot be zero, because B is the left inverse of A – we exclude the trivial case where $\mathbb{X} = \{0\}$.

The open mapping theorem provides the following elegant characterization of operators bounded below in Banach spaces.

15.6 Theorem *A continuous linear operator A mapping a Banach space $\mathbb{X}$ into a Banach space $\mathbb{Y}$ is bounded below if and only if it is injective and its range is closed.*[6]

Proof *(Necessity)* Injectivity is clear, and we focus on the closedness of the range. Hence, let $x_n, n \ge 1$ belong to $\mathbb{X}$ and suppose that $y := \lim_{n\to\infty} Ax_n$ exists. Condition (15.14) implies

$$\|x_n - x_m\| \le \delta^{-1}\|Ax_n - Ax_m\|, \qquad n,m \ge 1,$$

and this shows that $(x_n)_{n\ge 1}$ is a Cauchy sequence, because so is $(Ax_n)_{n\ge 1}$. For $x := \lim_{n\to\infty} x_n$ we have $\lim_{n\to\infty} Ax_n = Ax$, because A is continuous, and we infer that $y = Ax$ belongs to the range of A.

(Sufficiency) Let $\mathbb{Y}_0 \subset \mathbb{Y}$ denote the range of A. By assumption, $\mathbb{Y}_0$ is a closed subspace of $\mathbb{Y}$ and thus a Banach space itself. Moreover, $A\colon \mathbb{X} \to \mathbb{Y}_0$ is one-to-one and onto. Since both $\mathbb{X}$ and $\mathbb{Y}_0$ are Banach spaces the inverse operator A^{-1} is bounded by Corollary 15.5. We have seen that this implies (15.14) with $\delta = (\|A^{-1}\|_{\mathcal{L}(\mathbb{Y}_0,\mathbb{X})})^{-1}$. □

The above theorem can be summarized as follows: if we restrict ourselves to continuous linear operators in Banach spaces then an operator is bounded below if and only if it establishes a one-to-one correspondence between elements of its domain and its closed range.

As we shall see later, there are operators that are bounded below yet not continuous – see, for example, the closely related notion of *dissipative* operators, and the simple example of Exercise 15.3.

15.8 There Are No Other Norms in $C(S)$ but the Supremum Norm

The title of this section is appealing if somewhat imprecise; of course, there are a myriad of norms in $C(S)$ one can come up with (see e.g., Section 3.3), and additional myriads we will never even think of. The point is that we do not have in mind just any norm. First of all, especially in this book, we restrict ourselves to the norms that make $C(S)$ a complete space; this requirement reduces the number of possibilities, but there are still scores of norms that fit into this,

[6] We again exclude the case where $\mathbb{X}$ is trivial.

nomen omen, category. Second, it would be nice for the norm we want to take into account to have the following property: if a sequence $(x_n)_{n\geq 1}$ of elements of $C(S)$ converges in the norm to an $x \in C(S)$, then $(x_n(s))_{n\geq 1}$ converges to $x(s)$ for all $s \in S$. (In other words, convergence in the norm implies pointwise convergence.) For the sake of this discussion, we will call such norms *natural*. It is our goal in this section to show that as long as we restrict ourselves to natural norms that make $C(S)$ complete, there are no other norms but those that are equivalent to the classic supremum norm.

We start by recalling (compare Section 3.3) that two norms, say $\|\cdot\|_1$ and $\|\cdot\|_2$, in a linear space $\mathbb{X}$ are said to be equivalent if there are positive constants m and M such that

$$m\|x\|_1 \leq \|x\|_2 \leq M\|x\|_1, \qquad x \in \mathbb{X}.$$

Moreover, two norms are said to be *compatible* if for any sequence $(x_n)_{n\geq 1}$ of elements of $\mathbb{X}$, conditions $\lim_{n\to\infty} \|x_n - x\|_1 = 0$ and $\lim_{n\to\infty} \|x_n - \widetilde{x}\|_2 = 0$ imply $x = \widetilde{x}$. It is clear that equivalent norms are compatible: in fact, convergence of $(x_n)_{n\geq 1}$ in one of two equivalent norms implies convergence of $(x_n)_{n\geq 1}$ to the same limit in the other. The converse, however, is not true (see Exercise 15.4).

Nevertheless, as the following theorem shows, for complete norms compatibility is the same as equivalence.

15.7 Theorem *Suppose that either of the norms $\|\cdot\|_1$ and $\|\cdot\|_2$ makes a linear space $\mathbb{X}$ complete. That is, both $(\mathbb{X}, \|\cdot\|_1)$ and $(\mathbb{X}, \|\cdot\|_2)$ are complete spaces. If the norms $\|\cdot\|_1$ and $\|\cdot\|_2$ are compatible, they are equivalent.*

Proof We consider yet another norm:

$$|||x||| := \|x\|_1 + \|x\|_2, \qquad x \in \mathbb{X},$$

and claim that $\mathbb{X}$ with this norm is a complete space. Indeed, if $(x_n)_{n\geq 1}$ is a Cauchy sequence in this norm, then it is a Cauchy sequence in either of the original norms also. Since $(\mathbb{X}, \|\cdot\|_1)$ and $(\mathbb{X}, \|\cdot\|_2)$ are complete, there are $x \in \mathbb{X}$ and $\widetilde{x} \in \mathbb{X}$ such that $\lim_{n\to\infty} \|x_n - x\|_1 = 0$ and $\lim_{n\to\infty} \|x_n - \widetilde{x}\|_2 = 0$, and compatibility renders $x = \widetilde{x}$. It follows thus that $\lim_{n\to\infty} |||x_n - x||| = 0$, that is, that $(x_n)_{n\geq 1}$ converges in the sense of $|||\cdot|||$.

Next, we think of the operator A given by $Ax = x$, mapping $(\mathbb{X}, |||\cdot|||)$ to $(\mathbb{X}, \|\cdot\|_1)$. Clearly,

$$\|Ax\|_1 = \|x\|_1 \leq \|x\|_1 + \|x\|_2 = |||x|||, \qquad x \in \mathbb{X},$$

that is, A is bounded. This operator is one-to-one and 'onto' with $A^{-1}x = x$, and the open mapping theorem says thus that A^{-1} is also bounded. Therefore, there is a constant M such that

$$|||x||| \le M\|x\|_1, \qquad x \in \mathbb{X},$$

implying

$$\|x\|_2 \le M\|x\|_1, \qquad x \in \mathbb{X}.$$

By symmetry, there is also an $\widetilde{M}$ such that

$$\|x\|_1 \le \widetilde{M}\|x\|_2, \qquad x \in \mathbb{X}.$$

This completes the proof. □

In light of this theorem, our statement about the norm in $C(S)$ becomes clear. For, all natural norms are compatible with the supremum norm, and thus, as long as they make $C(S)$ complete, they are in fact equivalent to the supremum norm. In other words, a norm in $C(S)$ that makes this space complete is either equivalent to the supremum norm or is not natural, that is, convergence in this norm does not imply pointwise convergence.

15.9 The Closed Graph Theorem

Interesting operators that are encountered in pure and applied mathematics are seldom bounded; this is the case, for example, with most differential operators. Instead, a number of them is closed. To recall, a linear operator A with domain $\mathcal{D}(A)$ contained in a normed linear space $\mathbb{X}$ and range in another normed space $\mathbb{Y}$ is said to be closed if for any sequence $(x_n)_{n\ge1}$ of elements of $\mathcal{D}(A)$, $x \in \mathbb{X}$ and $y \in \mathbb{Y}$, conditions

$$\lim_{n\to\infty} \|x_n - x\|_{\mathbb{X}} = 0 \qquad \text{and} \qquad \lim_{n\to\infty} \|Ax_n - y\|_{\mathbb{Y}} = 0$$

imply that x is a member of $\mathcal{D}(A)$ and $Ax = y$.

For example, consider $\mathbb{X} = \mathbb{Y}$ that are equal to the space $C[0,1]$ of continuous functions on the unit interval $[0,1]$, let $\mathcal{D}(A)$ be the subset of $x \in C[0,1]$ that are continuously differentiable on this interval (with right-hand and left-hand derivatives at 0 and 1, respectively), and let $Ax = x'$. This operator cannot be bounded (more precisely: it cannot be extended to a bounded operator defined on the entire $C[0,1]$) because for x_n defined by $x_n(t) := t^n, t \in [0,1]$ we have $Ax_n = nx_{n-1}, n \ge 1$ and $\|x_n\| = 1$. However, A is closed.

For, suppose that functions $x_n \in \mathcal{D}(A)$ converge to an $x \in C[0,1]$, as $n \to \infty$, and at the same time x_n' converge to a $y \in C[0,1]$. Since each x_n is continuously differentiable,

$$x_n(t) = x_n(0) + \int_0^t x_n'(s)\, \mathrm{d}s, \qquad n \geq 1, t \in [0,1].$$

Letting $n \to \infty$ we obtain

$$x(t) = x(0) + \int_0^t y(s)\, \mathrm{d}s, \qquad t \in [0,1].$$

(We note that to obtain this conclusion it suffices to know, besides the convergence of $\left(x_n'\right)_{n\geq 1}$, that $\lim_{n\to\infty} x_n(0) = x(0)$; the assumption that $\lim_{n\to\infty} \|x_n - x\| = 0$ is superfluous.) This shows, however, that x is continuously differentiable with $x' = y$, and thus our claim is proven.

The situation presented above is typical in that a 'true' closed linear operator is never defined on the entire space. We make this statement precise in the following result, known as the closed graph theorem.

15.8 Theorem *Suppose $\mathbb{X}$ and $\mathbb{Y}$ are Banach spaces and $A\colon \mathbb{X} \to \mathbb{Y}$ is a linear map. (This means in particular that A is defined on the entire $\mathbb{X}$.) Then A is bounded whenever it is closed.*

Proof We need to show that in this case closedness implies continuity, the other implication being trivial. To this end, we consider a new norm in $\mathbb{X}$ defined by

$$\|x\|_* = \|x\|_{\mathbb{X}} + \|Ax\|_{\mathbb{Y}}, \qquad x \in \mathbb{X}.$$

(This norm is often referred to as the *graph norm*.) I claim that $\mathbb{X}$ with this norm is a Banach space.

For the proof, let $(x_n)_{n\geq 1}$ be a Cauchy sequence in this norm. Then $(x_n)_{n\geq 1}$ is a Cauchy sequence in $\mathbb{X}$ (in the original $\|\cdot\|_{\mathbb{X}}$ norm) and $(Ax_n)_{n\geq 1}$ is a Cauchy sequence in $\mathbb{Y}$. Therefore, there are $x \in \mathbb{X}$ and $y \in \mathbb{Y}$ such that $\lim_{n\to\infty} \|x_n - x\|_{\mathbb{X}} = 0$ and $\lim_{n\to\infty} \|Ax_n - y\|_{\mathbb{Y}}$. Since A is closed, x belongs to $\mathcal{D}(A)$, and $Ax = y$. It follows that $\lim_{n\to\infty} \|x_n - x\|_* = 0$, proving our claim.

The rest is immediate: the new norm is clearly compatible with the original norm (in fact, the new norm is no weaker than the original one, see Exercise 15.6), and thus – since either of them makes $\mathbb{X}$ a Banach space – is equivalent to the original norm. Hence, there is a constant $M > 0$ such that

$$\|x\|_{\mathbb{X}} + \|Ax\|_{\mathbb{Y}} \leq M\|x\|_{\mathbb{X}}, \qquad x \in \mathbb{X},$$

and this implies continuity of A. □

15.9 Example (Automatic continuity) As an illustration, consider a closed operator A in a Banach space $\mathbb{X}$, and assume that for a certain $\lambda \in \mathbb{R}$ there is precisely one solution x to the *resolvent equation*

$$\lambda x - Ax = y \tag{15.15}$$

regardless of which $y \in \mathbb{X}$ is chosen. We will argue that, by the closed graph theorem, the map $y \mapsto x$ is automatically bounded! The reader might recall that when discussing the existence of the right and left inverses we never bothered to comment on the continuity of these operators, restricting ourselves to saying that they correspond to the existence and uniqueness of solutions, respectively. The reason for that is the automatic continuity stated above.

By assumption, the operator, say B, that assigns the unique solution of equation (15.15) to the y on its right-hand side is well defined on the entire $\mathbb{X}$. By the closed graph theorem, it suffices to check that B is closed. To this end, assume that $(y_n)_{n\geq 1}$ converges to a y and at the same time that $(x_n)_{n\geq 1}$, where $x_n := By_n$, converges to an x. Then $(Ax_n)_{n\geq 1} = \lambda(x_n)_{n\geq 1} - (y_n)_{n\geq 1}$ converges also, namely, to $\lambda x - y$ and, since A is closed, we infer that $x \in \mathcal{D}(A)$ and $Ax = \lambda x - y$. But this means that $y \in \mathcal{D}(B)$ and $By = x$, as claimed.

For a yet more concrete illustration of automatic continuity, see Exercise 15.18.

15.10 Example (Automatic closedness) The previous example has a converse. Namely, assuming that for all y there is a unique x solving the resolvent equation (15.15) and that the map $y \mapsto x$ is continuous, we conclude that A needs to be closed.

Indeed, suppose that the sequence $(x_n)_{n\geq 1}$ of elements of $\mathcal{D}(A)$ converges to an x, and that $(Ax_n)_{n\geq 1}$ converges to a y. Also, let B be the map that assigns the solution to the resolvent equation to its right-hand side. Since B is continuous, and $(\lambda x_n - Ax_n)_{n\geq 1}$ converges to $\lambda x - y$, we have $\lim_{n\to\infty} x_n = \lim_{n\to\infty} B(\lambda x_n - Ax_n) = B(\lambda x - y)$. This means that x belongs to $\mathcal{D}(A)$, and $\lambda x - Ax = \lambda x - y$, that is, $Ax = y$, proving our claim. □

The reader has probably already noticed that both examples presented above come under one category. The real reason why these statements are true is as follows. An operator is closed if its graph is closed in the Cartesian product. Moreover, closedness of the graph does not change whether we look at x's as arguments and y's as values, or vice versa.

15.11 Example (Dissipative operators) Let $\mathbb{H}$ be a Hilbert space. An operator $A\colon \mathbb{H} \supset \mathcal{D}(A) \to \mathbb{H}$ is said to be *dissipative* if $(Ax, x) \leq 0$ for all $x \in \mathcal{D}(A)$. It is called *maximal dissipative* if the range of $I - A$ is the entire $\mathbb{H}$, that is, for any $y \in \mathbb{H}$ there is an $x \in \mathcal{D}(A)$ such that

$$x - Ax = y. \tag{15.16}$$

We will argue that maximal dissipative operators are necessarily *densely defined* (i.e., their domains are dense) and closed.

To prove the first claim we suppose that $\mathbb{H}_1 := \overline{\mathcal{D}(A)}$, the closure of $\mathcal{D}(A)$, is not the entire space $\mathbb{H}$ so that there is a $y_0 \in \mathbb{H}$ that does not belong to $\mathbb{H}_1$. Since $\mathbb{H}_1$ is a proper closed subspace of $\mathbb{H}$, there is the projection operator P that maps $\mathbb{H}$ onto $\mathbb{H}_1$, and we know that the vector $y := y_0 - Py_0$ is orthogonal to $\mathbb{H}_1$. However, by assumption, there is an $x \in \mathcal{D}(A)$ solving the resolvent equation (15.16). Multiplying both sides of this equation by x, we obtain $\|x\|^2 - (Ax, x) = (y, x) = 0$ because y is perpendicular to $\mathbb{H}_1$ and x belongs to $\mathcal{D}(A) \subset \mathbb{H}_1$. Now, since A is dissipative, we cannot have $(Ax, x) = \|x\|^2$ unless $x = 0$. But this implies that also $y = 0$, which is impossible, for by assumption y_0 does not belong to $\mathbb{H}_1$. This contradiction completes the proof of the first claim.

To prove the second, by Example 15.10, it suffices to show that (15.16) has precisely one solution x for any y and that the map that assigns x to y is a contraction. Now, if there are two solutions, say, x' and x'', of the resolvent equation for a y, then for $x := x' - x''$ we have $x = Ax$. Multiplying both sides of this relation by x we obtain $\|x\|^2 = (Ax, x)$, and we already established that this can only hold if $x = 0$. It follows that necessarily $x' = x''$, proving the uniqueness of solutions to the resolvent equation. Finally, multiplying (15.16) by x we see that

$$\|x\|^2 = (y, x) + (Ax, x) \le (x, y) \le \|x\| \|y\|,$$

and this shows that $\|x\| \le \|y\|$, as claimed.

15.12 Example (Operators associated with bilinear forms) It is often the case that the maximality of a given operator is much more difficult to prove than its dissipativity. Fortunately, condition (15.16) is automatically satisfied for dissipative operators associated with (non-negative, closed, symmetric, and densely defined) *bilinear forms*.

To explain the latter notion: suppose $\mathcal{D}$ is a dense linear subspace of a Hilbert space $\mathbb{H}$. A map $\mathcal{E} \colon \mathcal{D} \times \mathcal{D} \to \mathbb{R}$ is said to be a non-negative, symmetric bilinear form if

$$\mathcal{E}(x, y) = \mathcal{E}(y, x), \mathcal{E}(\alpha x + \beta y, z) = \alpha \mathcal{E}(x, z) + \beta \mathcal{E}(y, z) \text{ and } \mathcal{E}(x, x) \ge 0$$

for all $x, y, z \in \mathcal{D}$ and $\alpha, \beta \in \mathbb{R}$. Such a map induces a natural scalar product in $\mathcal{D}$, defined by

$$(x, y)_{\mathcal{D}} := \mathcal{E}(x, y) + (x, y), \qquad x, y \in \mathcal{D},$$

where, of course, $(\cdot, \cdot)$ denotes the scalar product in $\mathbb{H}$. If $\mathcal{D}$ with $(\cdot, \cdot)_{\mathcal{D}}$ is a Hilbert space, we say that $\mathcal{E}$ is closed.

With each closed symmetric and positive bilinear form there is an associated operator, say A. Its domain is composed of $x \in \mathcal{D}$ such that the map $\mathcal{D} \ni z \mapsto \mathcal{E}(x,z)$ can be extended to a bounded linear functional on the entire $\mathbb{H}$. In other words (see Section 12.4) x belongs to $\mathcal{D}(A)$ if and only if there is a $y \in \mathbb{H}$ such that $\mathcal{E}(x,z) = (y,z)$ for all $z \in \mathcal{D}$ (such a y is unique, $\mathcal{D}$ being dense in $\mathbb{H}$), and then

$$Ax = -y \qquad \text{so that} \qquad \mathcal{E}(x,z) = -(Ax,z), \quad z \in \mathcal{D}.$$

It is clear that the so-defined A is a linear operator. Moreover, it is dissipative: for $x \in \mathcal{D}(A)$, $(Ax,x) = -\mathcal{E}(x,x) \le 0$. To check that A is maximal, we fix a $y \in \mathbb{H}$ and consider the linear functional $\mathcal{D} \ni z \mapsto (y,z)$. This functional is bounded:

$$|(y,z)| \le \|y\|_{\mathbb{H}}\, \|z\|_{\mathbb{H}} \le \|y\|_{\mathbb{H}}\, \|z\|_{\mathcal{D}}$$

because $\|z\|^2_{\mathcal{D}} = \|z\|^2_{\mathbb{H}} + \mathcal{E}(z,z) \ge \|z\|^2_{\mathbb{H}}$. Hence, by the Riesz representation theorem, there is an $x \in \mathcal{D}$ such that $(y,z) = (x,z)_{\mathcal{D}} = (x,z) + \mathcal{E}(x,z)$ for $z \in \mathcal{D}$. The resulting relation $\mathcal{E}(x,z) = (y - x, z), z \in \mathcal{D}$ tells us that x is a member of $\mathcal{D}(A)$ and that $Ax = x - y$. This shows that A is maximal, as claimed.

For a simple example of the operator associated with the bilinear form, let $\mathbb{H} = \ell^2$ and $\mathcal{D} \subset \ell^2$ be composed of $(\xi_i)_{i\ge1}$ such that $\sum_{i=1}^\infty a_i\xi_i^2 < \infty$, where $(a_i)_{i\ge1}$ is a given sequence of non-negative numbers. $\mathcal{D}$ is dense in ℓ^2 because it contains all sequences that have only a finite number of non-zero coordinates. Letting

$$\mathcal{E}(x,y) = \sum_{i=1}^\infty a_i\xi_i\eta_i \tag{15.17}$$

for $x = (\xi_i)_{i\ge1}$ and $y = (\eta_i)_{i\ge1}$ in $\mathcal{D}$, we see that $\mathcal{E}$ is bilinear and symmetric; it is also non-negative since the a_i's are non-negative. The norm in $\mathcal{D}$ that is induced by the natural scalar product is given by $\|x\|^2_{\mathcal{D}} = \sum_{i=1}^\infty (1 + a_i)\xi_i^2$, and we know from Exercise 9.9 that $\mathcal{D}$ equipped with this norm is complete. In other words, $\mathcal{E}$ is closed.

We claim that the associated operator A is given by $A(\xi_i)_{i\ge1} = (-a_i\xi_i)_{i\ge1}$ with domain $\mathcal{D}(A)$ composed of $(\xi_i)_{i\ge1}$ such that $(-a_i\xi_i)_{i\ge1}$ belongs to ℓ^2. Indeed, for $x \in \mathcal{D}(A)$ and $z = (\zeta_i)_{i\ge1} \in \mathcal{D}$,

$$\mathcal{E}(x,z) = \sum_{i=1}^\infty a_i\xi_i\zeta_i = -\sum_{i=1}^\infty (-a_i\xi_i)\zeta_i = -(Ax,z),$$

where $(\cdot,\cdot)$ is the scalar product in ℓ^2. Conversely, suppose that an $x = (\xi_i)_{i\ge1}$ belongs to the domain of the operator associated with $\mathcal{E}$. Then, there is a $y =$

$(\eta_i)_{i\geq 1} \in \ell^2$ such that $\mathcal{E}(x,z) = (y,z), z \in \mathcal{D}$. For $z = e_n$ of (10.1), this relation renders $\eta_n = a_n\xi_n$, and so, n being arbitrary, $y = (a_i\xi_i)_{i\geq 1}$. Since y belongs to ℓ^2, we must have $\sum_{i=1}^{\infty} a_i^2\xi_i^2 < \infty$, that is, $x \in \mathcal{D}(A)$ and $y = -Ax$, completing the proof.

15.13 Example Here is a somewhat more involved example of the operator associated with a bilinear form (see [24] p. 327 for its full version). Let the a_i's in (15.17) be such that $\sum_{i=1}^{\infty} \frac{1}{1+a_i} < \infty$. Then, for $x = (\xi_i)_{i\geq 1} \in \mathcal{D}$, the series $\sum_{i=1}^{\infty} \xi_i$ converges absolutely. Indeed, by the Cauchy–Schwarz inequality,

$$\sum_{i=1}^{\infty} |\xi_i| = \sum_{i=1}^{\infty} \frac{1}{\sqrt{1+a_i}}\sqrt{1+a_i}|\xi_i| \leq \sqrt{\sum_{i=1}^{\infty} \frac{1}{1+a_i}}\sqrt{\sum_{i=1}^{\infty}(1+a_i)|\xi_i|^2}.$$

Since the second factor on the right-hand side above is $\|x\|_{\mathcal{D}}$, this estimate shows in fact that the map $(\xi_i)_{i\geq 1} \mapsto F(\xi_i)_{i\geq 1} := \sum_{i=1}^{\infty} \xi_i$ is a bounded linear functional in the Hilbert space $\mathcal{D}$. Hence, the kernel $\mathcal{D}_0 := \ker F := \{x \in \mathcal{D} \colon Fx = 0\}$ is a closed linear subspace of $\mathcal{D}$, and thus a Hilbert space itself.

It follows that $\mathcal{E}$ of (15.17), as restricted to $\mathcal{D}_0$, is a closed, non-negative bilinear form. Moreover, $\mathcal{D}_0$ is still dense in ℓ^2. To prove this, as in Example 15.11, it suffices to show that, except for $y = 0$, there are no other vectors that are perpendicular to $\mathcal{D}_0$. Now, a $y = (\eta_i)_{i\geq 1}$ that is perpendicular to $\mathcal{D}_0$ is perpendicular to all vectors $e_1 - e_n \in \mathcal{D}_0, n \geq 2$, where the e_n's are defined in (10.1). This implies $\eta_n = \eta_1$. Such a y does not belong to ℓ^2 unless $\eta_1 = 0$, that is, $y = 0$.

We will show that an $x = (\xi_i)_{i\geq 1}$ belongs to the domain of the associated operator A if and only if there is a constant c such that $(c + a_i\xi_i)_{i\geq 1}$ belongs to ℓ^2, and then $Ax = -(c + a_i\xi_i)_{i\geq 1}$. (Such a constant is necessarily unique, because otherwise a non-zero constant sequence would belong to ℓ^2.) Indeed, if there is such a constant, for any $z = (\zeta_i)_{i\geq 1} \in \mathcal{D}_0$, we have

$$\mathcal{E}(x,z) = \sum_{i=1}^{\infty} a_i\xi_i\zeta_i = \sum_{i=1}^{\infty}(c+a_i\xi_i)\zeta_i - c\sum_{i=1}^{\infty}\zeta_i = \sum_{i=1}^{\infty}(c+a_i\xi_i)\zeta_i = -(Ax,z).$$

Conversely, assume that for an $x = (\xi_i)_{i\geq 1} \in \mathcal{D}_0$ one can find a $y = (\eta_i)_{i\geq 1} \in \ell^2$ such that $\mathcal{E}(x,z) = (y,z)$ for $z = (\zeta_i)_{i\geq 1} \in \mathcal{D}_0$, that is, $\sum_{i=1}^{\infty}(\eta_i - a_i\xi_i)\zeta_i = 0$. Taking $z = e_1 - e_n, n \geq 2$, as above, we obtain then $\eta_n - a_n\xi_n = \eta_1 - a_1\xi_1, n \geq 2$. Thus, there exists a $c \in \mathbb{R}$ such that $\eta_i = c + a_i\xi_i, i \geq 1$, and since y belongs to ℓ^2 so does $(c + a_i\xi_i)_{i\geq 1}$, as claimed.

15.10 Exercises

Exercise 15.1. Let $D_{1,\ell}$ be the subset of $C[0,1]$ defined as follows: an $x \in C[0,1]$ belongs to $D_{1,\ell}$ if for all $h \in (0,1]$, $|x(1-h) - h(1)| \le h\ell$. Argue as in Section 15.2 to see that each $D_{1,\ell}$ is closed and yet nowhere dense, and that $D_1 := \bigcup_{\ell \ge 1} D_{1,\ell}$ contains all functions in $C[0,1]$ that have finite left-hand derivatives at 1. Deduce, by noting that $\bigcup_{k\ge 1} D_k$ is meagre, that so is the set of functions that do have a finite derivative inside $[0,1]$ or a one-handed derivative at 0 or 1.

Exercise 15.2. The analysis presented in Section 14.4.2 shows that the partial sums of the Fourier series of an $x \in C_{\mathrm{p}}[-\pi,\pi]$ are given by

$$s_n(t) = \frac{1}{\pi}\int_{-\pi}^{\pi} x(t+s)\frac{\sin(n+\frac{1}{2})s}{2\sin\frac{s}{2}}\,\mathrm{d}s, \qquad t \in [-\pi,\pi], n \ge 0,$$

and from Section 15.4 we know that $(s_n(t))_{n\ge 1}$ converges on a rather small set of x's. Use the Riemann lemma (Exercise 7.7) to prove the following positive result: if x has a derivative at a $t \in [-\pi,\pi]$ (for $t = \pm\pi$ this means that the right-hand derivative at $-\pi$ and the left-hand derivative at π exist and are equal to each other) then $\lim_{n\to\infty} s_n(t) = x(t)$. **Hint:** Write $s_n(t) - s(t)$ as

$$\frac{1}{\pi}\int_{-\pi}^{\pi} \frac{x(t+s)-x(t)}{s}\,\frac{s}{2\sin\frac{s}{2}}\,\sin(n+\tfrac{1}{2})s\,\mathrm{d}s$$

and note that $[-\pi,\pi]\setminus\{0\} \ni s \mapsto \frac{x(t+s)-x(t)}{s}\frac{s}{2\sin\frac{s}{2}}$ can be extended to a continuous function on $[-\pi,\pi]$. Moreover, $\sin(n+\frac{1}{2})s = \sin ns\cos\frac{1}{2}s + \cos ns\sin\frac{1}{2}s$.

Exercise 15.3. Let c_{00} be the space of sequences $(\xi_i)_{i\ge 1}$ that have all but a finite number of non-zero coordinates; the norm in this space is $\|(\xi_i)_{i\ge 1}\| = \max_{i\ge 1}|\xi_i|$. Check to see that the map $A\colon c_{00} \to c_{00}$ given by $A(\xi_i)_{i\ge 1} = (i\xi_i)_{i\ge 1}$ is not bounded and yet bounded below.

Exercise 15.4. In ℓ^1 consider the norm

$$\|\,(\xi_i)_{i\ge 1}\,\| := \max_{i\ge 1}|\xi_i|.$$

Check to see that this norm is compatible with but not equivalent to the standard norm in ℓ^1. Note that ℓ^1 is not complete with this norm.

Exercise 15.5. Convince yourself that in ℓ^1 there is no other norm but $\|\,(\xi_i)_{i\ge 1}\,\| = \sum_{i\ge 1}|\xi_i|$ as long as we restrict ourselves to natural norms.

Exercise 15.6. A norm $\|\cdot\|_1$ in a linear space $\mathbb{X}$ is said to be no weaker than a norm $\|\cdot\|_2$ if for all sequences $(x_n)_{n\geq 1}$ of elements of $\mathbb{X}$, condition $\lim_{n\to\infty}\|x_n\|_1 = 0$ implies $\lim_{n\to\infty}\|x_n\|_2$. Check to see that this definition is equivalent to the existence of a $C > 0$ such that $\|x\|_2 \leq C\|x\|_1$ for all $x \in \mathbb{X}$. **Hint:** If there is a sequence $(x_n)_{n\geq 1}$ such that $\|x_n\|_2 > n\|x_n\|_1 \neq 0$, consider $\widetilde{x}_n := \frac{1}{\sqrt{n}\|x_n\|_1}x_n$.

Exercise 15.7. Let $\mathbb{X}$ be a Banach space, and $\mathbb{Y}$ be a normed space. Suppose also that an operator $A\colon \mathbb{X} \supset \mathcal{D}(A) \to \mathbb{Y}$ is such that $\|Ax\|_{\mathbb{Y}} \geq \delta\|x\|_{\mathbb{X}}$ for a certain $\delta > 0$ and all $x \in \mathcal{D}(A)$. Check to see that A is closed if and only if its range is closed. A **hint** can be found in the proof of Theorem 15.6.

Exercise 15.8. Let $\mathbb{X}$ be a Banach space, and let $A\colon \mathcal{D}(A) \to \mathbb{X}$ be a closed operator. Moreover, let $B \in \mathcal{L}(\mathbb{X})$ be such that $Bx \in \mathcal{D}(A)$ for each $x \in \mathbb{X}$. Then, the composition $AB\colon \mathbb{X} \to \mathbb{X}$ is well defined. Show that this operator is bounded. **Hint:** Show first that this operator is closed.

Exercise 15.9. Let A and B be closed linear operators in a Banach space with common domain $\mathcal{D}(A) = \mathcal{D}(B)$. Show that there is a constant $M > 0$ such that

$$M^{-1}(\|x\| + \|Ax\|) \leq \|x\| + \|Bx\| \leq M(\|x\| + \|Ax\|), \qquad x \in \mathcal{D}(A).$$

Hint: Norms $\|\cdot\|_A$ and $\|\cdot\|_B$ defined by $\|x\|_A := \|x\| + \|Ax\|$ and $\|x\|_B := \|x\| + \|Bx\|$, $x \in \mathcal{D}(A)$, are compatible.

Exercise 15.10. ▲ An operator B is said to *extend* an operator A if $\mathcal{D}(A) \subset \mathcal{D}(B)$ and $Bx = Ax$ for $x \in \mathcal{D}(A)$. An operator A is said to be *closable* if conditions $x_n \in \mathcal{D}(A)$, $\lim_{n\to\infty} x_n = 0$ and $\lim_{n\to\infty} Ax_n = y$ imply $y = 0$. Prove that A is closable if and only if there is a closed operator B that extends A. **Hint:** Sufficiency of the existence of B is simple; for the necessity part define $\overline{A}x = \lim_{n\to\infty} Ax_n$ whenever there are $x_n \in \mathcal{D}(A)$ such that $\lim_{n\to\infty} x_n = x \in \mathbb{X}$ and $\lim_{n\to\infty} Ax_n$ exists (this definition does not depend on the choice of $(x_n)_{n\geq 1}$); check that the so-defined $\overline{A}$ is closed. ($\overline{A}$ is referred to as the *closure of* A.)

Exercise 15.11. Let $\mathbb{X}$ be a Banach space, let A in $\mathbb{X}$ be closable, and assume that B is bounded and maps $\mathbb{X}$ into $\mathcal{D}(A)$. Prove that AB is closed, and thus continuous.

Exercise 15.12. Let g be a given function in $C[0, 1]$ such that $g(1) = 0$. Check that for any $c \in \mathbb{R}$, $\|\cdot\|_c$ given by $\|f\|_c = \|f - cf(1)g\|$, $f \in C[0, 1]$ is a norm in $C[0, 1]$ that is equivalent to the usual supremum norm. This can be proved with or without referring to Exercise 13.9 and with or without referring to Section 15.8.

Exercise 15.13. Suppose the resolvent equation (15.15) has a unique solution x for all $y \in \mathbb{X}$, provided that λ belongs to a set $\Lambda \subset \mathbb{R}$. Denote the solution corresponding to λ and y by $R_\lambda y$. Prove the following Hilbert equation:

$$R_\lambda - R_\mu = (\mu - \lambda) R_\lambda R_\mu, \qquad \lambda, \mu \in \Lambda.$$

Deduce in particular that the operators $R_\lambda, \lambda \in \Lambda$ commute. **Hint:** Since $R_\lambda R_\mu y$ is a unique solution to $\lambda x - Ax = R_\mu y$, it suffices to check that $x' := \frac{1}{\mu - \lambda}(R_\lambda y - R_\mu y)$ solves this equation also.

Exercise 15.14.

(a) In the space $C[0,1]$ of real continuous functions on the unit interval we define the family of operators $\{T(t), t \geq 0\}$ by

$$T(t)x(s) = x(\mathrm{e}^{-at}s), \qquad x \in C[0,1], t \geq 0, s \in [0,1],$$

where $a \geq 0$ is a given number. Prove that $\{T(t), t \geq 0\}$ is a strongly continuous semigroup of contractions.

(b) Repeat the exercise for the space $C[0,\infty]$ of real continuous functions defined on the right half-axis that have finite limits at $+\infty$, and for the operators defined by

$$T(t)x(s) = x(s + at), \qquad x \in C[0,\infty], s \geq 0, t \geq 0, \qquad (15.18)$$

where again $a \geq 0$.

(c) Do the same in ℓ^2 for the family

$$T(t)(\xi_i)_{i \geq 1} = \left(\mathrm{e}^{-ait}\xi_i\right)_{i \geq 1}, \qquad (\xi_i)_{i \geq 1} \in \ell^2, t \geq 0;$$

$a \geq 0$ again being given.

Exercise 15.15. Use the argument of Section 14.6 to check that formula (15.18), when applied to members x of $L^1(\mathbb{R}^+)$, also defines a strongly continuous semigroup in this space.

Exercise 15.16. Let $\{S(t), t \geq 0\}$ be a strongly continuous semigroup of bounded linear operators in a Banach space $\mathbb{X}$, let and $B \in \mathcal{L}(\mathbb{X})$ be an operator such that $B^2 = B$ and $S(t)B = B, t \geq 0$. Prove that $\{T(t), t \geq 0\}$ defined by

$$T(t)x = S(t)x + B \int_0^t \mathrm{e}^{t-s} S(s)x \, \mathrm{d}s, \qquad x \in \mathbb{X}, t \geq 0,$$

is also a strongly continuous semigroup. (In Section 15.5 we have proved that $[0,\infty) \ni s \mapsto S(s)x \in \mathbb{X}$ is continuous (in the norm of $\mathbb{X}$). Hence, by the main result of Chapter 6, for each $t > 0$, $[0,t] \ni s \mapsto e^{t-s}S(s)x \in \mathbb{X}$ is Riemann integrable.)

Exercise 15.17. Let $\{S(t), t \geq 0\}$ be a strongly continuous semigroup of bounded linear operators in a Banach space $\mathbb{X}$, and let $B \in \mathcal{L}(\mathbb{X})$ be an operator such that $B^2 = bB$ for some $b \in \mathbb{R}$, and $BS(t) = B, t \geq 0$. Prove that $\{T(t), t \geq 0\}$ defined by

$$T(t)x = S(t)x + \int_0^t e^{b(t-s)}S(s)Bx\,ds, \qquad x \in \mathbb{X}, t \geq 0,$$

is also a strongly continuous semigroup.

Exercise 15.18. Let A be an operator in $C[0,1]$ with domain composed of continuously differentiable $x \in C[0,1]$ such that $x(1) = 0$, and let $Ax = x'$. Check that A is closed, and that for all $\lambda \in \mathbb{R}$, the resolvent equation $\lambda x - Ax = y$ has precisely one solution. The map $y \mapsto x$ is continuous, isn't it? **Hint:** The solution is given by $x(t) := \int_t^1 e^{\lambda(t-s)}y(s)\,ds, t \in [0,1]$.

Exercise 15.19. Prove the following generalization of Exercise (12.7): let A be a closed operator in a Banach space $\mathbb{X}$, and let $[a,b] \ni t \mapsto x(t) \in \mathbb{X}$ be a Riemann integrable function such that $[a,b] \ni t \mapsto Ax(t)$ is also integrable. Then

$$A\int_a^b x(t)\,dt = \int_a^b Ax(t)\,dt.$$

Exercise 15.20. Repeat Example 15.13 with the functional F redefined by the formula $F(\xi_i)_{i\geq 1} = \sum_{i=1}^{\infty}(-1)^i\xi_i, (\xi_i)_{i\geq 1} \in \mathcal{D}$.

Exercise 15.21. Let $L^2[a,1]$ be the space of square integrable real functions defined on the interval $[a,1]$, where $a \in (0,1)$. Also, let $\mathcal{D} \subset L^2[a,1]$ be the subspace of functions that are absolutely continuous in $[a,1]$ with generalized derivatives in $L^2[a,1]$ (see Section 11.3). For $x, y \in \mathcal{D}$, let

$$\mathcal{E}(x,y) = abx(a)y(a) + \int_a^1 sx'(s)y'(s)\,ds,$$

where $b > 0$ is a given constant. (a) Show that $\mathcal{E}$ is a non-negative, closed and symmetric bilinear form. (b) Taking for granted that $\mathcal{D}$ is dense in $L^2[a,1]$, characterize the operator associated with $\mathcal{E}$. **Hint for (a):** If $(x_n)_{n\geq 1}$ is a Cauchy sequence in $\mathcal{D}$, there are $x, y \in L^2[a,1]$ such that $\lim_{n\to\infty} x_n = x$ and $\lim_{n\to\infty} x_n' = y$ in the norm of this space, whereas the numerical sequence $(x_n(a))_{n\geq 1}$ converges to a $C \in \mathbb{R}$. It follows that $x(s) = C + \int_a^s y(u)\,du, s \in [a,1]$, and so $\lim_{n\to\infty} \|x_n - x\|_{\mathcal{D}} = 0$. **Hint for (b):** An $x \in \mathcal{D}$ belongs to $\mathcal{D}(A)$ if and only if (i) $s \mapsto sx'(s)$ is absolutely continuous with $s \mapsto (sx'(s))'$

in $L^2[a,1]$, and (ii) $x'(1) = 0$ and $x'(a) = bx(a)$. Then $Ax(s) = (sx'(s))', s \in [a,1]$. To prove this assertion use the integration by parts formula, which works with generalized derivatives as it does with classical derivatives. In showing that an $x \in \mathcal{D}(A)$ must satisfy conditions (i) and (ii), you will encounter the formula

$$abx(a)z(a) + \int_a^1 sx'(s)z'(s)\,\mathrm{d}s = z(1)\int_a^1 y(s)\,\mathrm{d}s - \int_a^1 \int_a^s y(u)\,\mathrm{d}u\, z'(s)\,\mathrm{d}s,$$

which is to hold for all $z \in \mathcal{D}$. Note then that a $z \in \mathcal{D}$ satisfies $z(a) = z(1) = 0$ if and only if there is a $w \in L^2[a,b]$ such that $\int_a^1 w(s)\,\mathrm{d}s = 0$ and $z(s) = \int_a^s w(u)\,\mathrm{d}u, s \in [a,1]$. For such a z, the formula takes the form $\int_a^1 sx'(s)w(s)\,\mathrm{d}s = -\int_a^1 \int_a^s y(u)\,\mathrm{d}u\, w(s)\,\mathrm{d}s$ and proves that $s \mapsto sx'(s) + \int_a^s y(u)\,\mathrm{d}u$ is perpendicular to all w's. Since the w's form the subspace that is perpendicular (cf. Exercise 9.11) to the subspace of constant functions, there is a C such that $sx'(s) = C - \int_a^s y(u)\,\mathrm{d}u, s \in [a,1]$. Hence, on the one hand, $C = ax'(a)$, and, on the other, $abx(a)z(a) + C(z(1) - z(a)) = z(1)\int_a^1 y(s)\,\mathrm{d}s$ for all $z \in \mathcal{D}$.

☞ CHAPTER SUMMARY

The celebrated Baire category theorem says that a complete space cannot be represented as a countable union of nowhere dense sets. This is a fundamental description of the structure of complete spaces. Because of that, it is fitting to derive the Banach–Steinhaus theorem as a consequence of Baire's. This is what we do at the beginning of this chapter. We also show that the set of differentiable functions is quite small (i.e., meagre) in the space of continuous functions. As further consequences of Baire's theorem we discuss two other fundamental results of functional analysis, the open mapping theorem and the closed graph theorem, together with some of their most immediate applications. In the meantime, we use the Banach–Steinhaus theorem to show that a Fourier series cannot converge uniformly for all continuous (and periodic) functions.

16
Semigroups of Operators

Many important equations of mathematical physics and mathematical biology have the form (13.12), but with unbounded A. In fact, Theorem 13.7 can be thought of as a negative result. For, it demonstrates that, in the case of a bounded operator, the initial value determines the future and the past states of the described process. This is contrary to our intuition, which says that, if the present state is capable of determining anything at all, it could only determine the future, for the present state could have been obtained by many possible pasts. Hence, either the equation has to involve an unbounded operator or at least, if its solutions can be extended into the past, their parts corresponding to negative times should have no clear 'physical' interpretation (as it was in the case of Example 13.10).

This necessitates an analysis of operators A for which a counterpart of Theorem 13.7 holds: in this new theorem the initial value should determine a unique solution merely for $t \geq 0$, not for all $t \in \mathbb{R}$. In other words, we consider the Cauchy problem (initial value problem) of the form

$$u'(t) = Au(t), \quad u(0) = x \in \mathbb{X}, \qquad t \geq 0. \tag{16.1}$$

If we want solutions to this equation to be unique and to depend continuously on initial data x, we will be led to the notion of the strongly continuous semigroup, introduced in Section 15.5. Because of the semigroup property (15.8), semigroups are akin to exponential functions, and in many aspects mimic their properties.

In this chapter, we present the rudiments of the theory of semigroups of operators, and in particular discuss its fundamental pillar – the generation theorem of Hille–Yosida–Phillips–Feller–Miyadera. The reader of this book will not fail to notice that the proof of the theorem in question relies heavily on the completeness of the space we work in.

Figure 16.1 The theory of semigroups is the art of assigning exponential functions to operators (original image: Greg Pease/Getty Images)

16.1 Generators of Semigroups

We start by introducing the crucial notion of a semigroup's *generator*. The entire theory can in fact be seen as the art of assigning 'exponential functions' to generators (see Figure 16.1), and the Hille–Yosida–Phillips–Feller–Miyadera theorem tells us to which operators one can associate 'exponential functions,' that is, semigroups, and to which one cannot.

Given a strongly continuous semigroup $\{T(t), t \ge 0\}$ in a Banach space $\mathbb{X}$, we define its generator, say A, as the following operator, $A\colon \mathcal{D}(A) \to \mathbb{X}$:

$$Ax = \lim_{h \to 0+} h^{-1}(T(h)x - x) \tag{16.2}$$

with domain $\mathcal{D}(A) \subset \mathbb{X}$ composed of x's such that the limit on the right-hand side above exists (in the norm of $\mathbb{X}$). Thus, Ax is the right-hand derivative of the $\mathbb{X}$-valued function $t \mapsto T(t)x$ at $t = 0$.

It is important to note here that, although the so-defined A is clearly linear, it is seldom bounded, but possesses the following four important properties:

1. Its domain is invariant under $\{T(t), t \ge 0\}$: that is, $T(t)x \in \mathcal{D}(A)$ for all $t \ge 0$ whenever $x \in \mathcal{D}(A)$, and we have $AT(t)x = T(t)Ax$.

2. For x in $\mathcal{D}(A)$, the function $t \mapsto T(t)x$ (called the semigroup's *trajectory*) is differentiable at all $t \geq 0$, with $\frac{\mathrm{d}}{\mathrm{d}t}T(t)x = AT(t)x = T(t)Ax$.
3. A is densely defined.
4. A is closed.

The first property is easy to prove with the help of the semigroup relation (15.8): for $y := T(t)x$,

$$h^{-1}(T(h)y - y) = h^{-1}(T(t+h)x - T(t)x) = T(t)h^{-1}(T(h)x - x); \quad (16.3)$$

and since $h^{-1}(T(h)x - x)$ converges to Ax as $h \to 0+$ and $T(t)$ is a bounded operator, we see that the entire expression converges to $T(t)Ax$. This demonstrates property 1.

Turning to property 2., we note first that it is an analogue of the fact that all semigroup trajectories are continuous functions, and that its proof goes along the lines presented in Section 15.5. Second, we have already proved 'half of' 2.: the second expression in (16.3) is the right-hand difference quotient of the trajectory we are studying, and we proved that it converges to $T(t)Ax$, that is, that the right-hand derivative of the trajectory exists and equals $T(t)Ax$. Hence, we are left with showing that the same is true of the left-hand derivative. Now, the left-hand difference quotient at $t > 0$,

$$-h^{-1}(T(t-h)x - T(t)x) = -T(t-h)h^{-1}(x - T(h)) = T(t-h)h^{-1}(T(h)x - x),$$

indeed converges to $T(t)Ax$, as $h \to 0+$. For, given $\epsilon > 0$ we can find $h_0 < t$ so small that simulateneously

$$\|h^{-1}(T(h)x - x) - Ax\| < \epsilon \qquad \text{and} \qquad \|T(t-h)Ax - T(t)Ax\| < \epsilon$$

for $0 < h < h_0$, and then, for M and ω introduced in (15.9),

$$\begin{aligned} \|T(t-h)h^{-1}(T(h)x - x) - T(t)Ax\| &\leq \|T(t-h)[h^{-1}(T(h)x - x) - Ax]\| \\ &\quad + \|T(t-h)Ax - T(t)Ax\| \\ &< \|T(t-h)\|\|h^{-1}[T(h)x - x] - Ax\| + \epsilon \\ &\leq (M\mathrm{e}^{(t-h)\omega} + 1)\epsilon \leq (M\mathrm{e}^{t|\omega|} + 1)\epsilon. \end{aligned}$$

To prove property 3., we recall that, for any $x \in \mathbb{X}$, the trajectory $t \mapsto T(t)x$ is continuous. It follows that for any $0 \leq a \leq b$, the Riemann integral $\int_a^b T(s)x \,\mathrm{d}s$ is well defined (see Chapter 6). Moreover, by (6.5),

$$\lim_{h \to 0+} h^{-1} \int_t^{t+h} T(s)x \,\mathrm{d}s = T(t)x, \qquad \text{for all } t \geq 0, x \in \mathbb{X}. \quad (16.4)$$

This, in turn, shows that, for any $x \in \mathbb{X}$ and $t > 0$, $y := \int_0^t T(s)x\,\mathrm{d}s$ belongs to $\mathcal{D}(A)$. Indeed,

$$\begin{aligned} T(h)y - y &= \int_0^t T(h+s)x\,\mathrm{d}s - \int_0^t T(s)x\,\mathrm{d}s = \int_h^{t+h} T(s)x\,\mathrm{d}s - \int_0^t T(s)x\,\mathrm{d}s \\ &= \int_t^{t+h} T(h+s)x\,\mathrm{d}s - \int_0^h T(s)x\,\mathrm{d}s \end{aligned}$$

provided that $h < t$. This establishes, by (16.4), that $\lim_{h\to 0+} h^{-1}(T(h)y - y) = T(t)x - x$, that is, that $\int_0^t T(s)x\,\mathrm{d}s$ belongs to $\mathcal{D}(A)$ (as claimed) and

$$A \int_0^t T(s)x\,\mathrm{d}s = T(t)x - x. \tag{16.5}$$

Since $t^{-1}\int_0^t T(s)x\,\mathrm{d}s$ belongs to $\mathcal{D}(A)$ whenever $\int_0^t T(s)x\,\mathrm{d}s$ does and, again by (16.4), $\lim_{t\to 0+} t^{-1}\int_0^t T(s)x\,\mathrm{d}s = x$, we conclude thus that $\mathcal{D}(A)$ is indeed dense in $\mathbb{X}$, $x \in \mathbb{X}$ being arbitrary.

We are therefore left with demonstrating property 4. As a preparation, we recall that for $x \in \mathcal{D}(A)$, the trajectory $t \mapsto T(t)x$ is differentiable with derivative $t \mapsto T(t)Ax$ (by the already proven point 2.). Hence, the second fundamental theorem of calculus (see Chapter 6) renders the following counterpart of (16.5):

$$\int_0^t T(s)Ax\,\mathrm{d}s = T(t)x - x, \qquad t \geq 0. \tag{16.6}$$

With this relation under our belt, we let $x_n, n \geq 1$ be members of $\mathcal{D}(A)$ such that the limit $\lim_{n\to\infty} x_n$ exists and equals $x \in \mathbb{X}$, and at the same time $\lim_{n\to\infty} Ax_n = y$ for some $y \in \mathbb{X}$. We have thus

$$\int_0^t T(s)Ax_n\,\mathrm{d}s = T(t)x_n - x_n, \qquad n \geq 1, t \geq 0. \tag{16.7}$$

Since $\lim_{n\to\infty} A_n x = y$, for fixed $t > 0$, the integrand above converges uniformly with respect to $s \in [0,t]$ to the trajectory $s \mapsto T(s)y$, and this implies that the left-hand side converges to $\int_0^t T(s)y\,\mathrm{d}s$. Since, at the same time, the right-hand side converges to $T(t)x - x$, it follows that

$$\int_0^t T(s)y\,\mathrm{d}s = T(t)x - x, \qquad t \geq 0.$$

Hence, we obtain $\lim_{t\to 0+} t^{-1}(T(t)x - x) = y$ as a special case of (16.4). This, however, means that $x \in \mathcal{D}(A)$ and $Ax = y$, as desired.

Before closing this section, we note that conditions 1. through 3. can be summarized as follows: for $x \in \mathcal{D}(A)$, that is, for x in a dense set, $t \mapsto u(t) =$

$T(t)x$ is continuously differentiable, $u(t)$ belongs to $\mathcal{D}(A)$ for all $t \geq 0$, and we have

$$\frac{\mathrm{d}u(t)}{\mathrm{d}t} = Au(t), \qquad t \geq 0;$$

also, $u(0) = x$. This, by definition, means that the trajectory of the semigroup $\{T(t), t \geq 0\}$ emanating from $x \in \mathcal{D}(A)$ is a *classical solution* to the Cauchy problem (16.1) with A equal to the semigroup's generator. (Conversely, it can be argued that the existence of (unique) classical solutions for all $x \in \mathcal{D}(A)$ plus their continuous dependence on initial values implies the existence of the semigroup generated by A.) Conversely, the trajectory starting at an $x \in \mathbb{X} \setminus \mathcal{D}(A)$ is not differentiable (at least not at $t = 0$) and thus cannot be the classical solution to the Cauchy problem. Nevertheless, in view of (16.5), such trajectories are said to be *mild solutions*.

16.2 Examples of Generators

In this section we will calculate explicitly generators of a couple of semigroups introduced in the exercises of Chapter 15. We start with the simplest of these, related to a semigroup in ℓ^2.

16.2.1 The Generator of the Semigroup of Exercise 15.14 (c)

Here and in the next example we exclude the trivial case of $a = 0$ in which $T(t)x = x, t \geq 0$ and thus the generator coincides with the zero operator.

We want to identify elements of $\mathcal{D}(A)$ and find a formula for $A(\xi_i)_{i\geq 1}$ for $(\xi_i)_{i\geq 1} \in \mathcal{D}(A)$. The idea is to deduce as many properties as we can of $(\xi_i)_{i\geq 1} \in \mathcal{D}(A)$ from the existence of the defining limit (16.2) in the hope that these properties characterize elements of $\mathcal{D}(A)$. In our case,

$$h^{-1}(T(h)(\xi_i)_{i\geq 1} - (\xi_i)_{i\geq 1}) = \left(h^{-1}(\mathrm{e}^{-aih} - 1)\xi_i\right)_{i\geq 1}, (\xi_i)_{i\geq 1} \in \ell^2, h > 0;$$

and, since convergence in the norm of ℓ^2 implies convergence of coordinates, we see that the limit as $h \to 0+$, if it does exist, must coincide with $a\,(-i\xi_i)_{i\geq 1}$. In other words, $a\,(-i\xi_i)_{i\geq 1}$ is the only, and natural, candidate for $A(\xi_i)_{i\geq 1}$, and since $A(\xi_i)_{i\geq 1}$, by definition, belongs to ℓ^2, we conclude that $\mathcal{D}(A)$ is contained in the subset of $(\xi_i)_{i\geq 1} \in \ell^2$ such that $(-i\xi_i)_{i\geq 1} \in \ell^2$.

So, the question is whether the requirement that $(-i\xi_i)_{i\geq 1}$ belongs to ℓ^2 suffices for $(\xi_i)_{i\geq 1}$ to belong to $\mathcal{D}(A)$. Fortunately, it does. Indeed, since $0 \leq h^{-1}(1 - \mathrm{e}^{-aih}) = h^{-1}\int_0^h ai\mathrm{e}^{-ais}\,\mathrm{d}s \leq ai$, the square of the norm of the difference between $h^{-1}(T(h)(\xi_i)_{i\ge1} - (\xi_i)_{i\ge1})$ and $a\,(-i\xi_i)_{i\ge1}$, that is, the square of the norm of the vector $\left((ai - h^{-1}\int_0^h ai\mathrm{e}^{-ais}\,\mathrm{d}s)\xi_i\right)_{i\ge1}$, equals

$$\sum_{i\ge1}(ai - h^{-1}\int_0^h ai\mathrm{e}^{-ais}\,\mathrm{d}s)^2\xi_i^2,$$

and this series is dominated by $4a^2\sum_{i\ge1} i^2\xi_i^2$. It follows that the sum above tends to zero as $h \to 0+$ because then each summand in the series converges to zero; this is a direct consequence of the Lebesgue dominated convergence theorem but, as in Section 14.1, can be worked out in an elementary manner.

We have thus proved that $(\xi_i)_{i\ge1} \in \ell^2$ belongs to the domain of the generator of the studied semigroup if and only if $(-i\xi_i)_{i\ge1}$ also belongs to ℓ^2, and then $A(\xi_i)_{i\ge1} = a\,(-i\xi_i)_{i\ge1}$.

16.2.2 The Generator of the Semigroup of Exercise 15.14 (a)

We argue along the same lines as in the previous example but use a special characteristic of the space $C[0,1]$ in which the semigroup of interest is defined. We note that if x belongs to the domain $\mathcal{D}(A)$ of the generator of this semigroup, the limit

$$\lim_{h\to0+} h^{-1}(x(\mathrm{e}^{-ah}s) - x(s)) = Ax(s) \tag{16.8}$$

exists for each $s \in [0,1]$ (in fact, this limit must be uniform in s). For $s = 0$ the left-hand side is 0, and so this condition says merely that $Ax(0) = 0$. For $s > 0$, however, in terms of $\Delta = \Delta(h) := s(1 - \mathrm{e}^{-ah}) > 0$ it says that $\lim_{h\to0+}\frac{\Delta}{h}\frac{x(s-\Delta)-x(s)}{\Delta}$ exists. Since $\lim_{h\to0+}\frac{\Delta}{h} = as$, we see that x has the left-hand derivative $x'_-(s)$ at all points $s \in (0,1]$. It follows that $Ax(s) = -asx'_-(s), s \in (0,1]$. However, Ax belongs to $C[0,1]$, and this implies that $(0,1] \ni s \mapsto x'_-(s)$ is continuous. A theorem of real analysis tells us thus that x is differentiable at all points $s \in (0,1]$, and so the formula for Ax takes the form

$$Ax(s) = -asx'(s), \qquad s \in (0,1].$$

Since Ax is a continuous function, we conclude our reasoning by saying that $\mathcal{D}(A)$ is contained in the set, say, $\mathcal{D}$, of $x \in C[0,1]$ that are continuously differentiable in $s \in (0,1]$ and such that $\lim_{s\to0+} sx'(s) = 0$.

To prove the reverse inclusion $\mathcal{D} \subset \mathcal{D}(A)$ we note first that for $s = 0$, the left-hand side of (16.8) is zero, and for $s > 0$ it equals $sh^{-1}(\mathrm{e}^{-ah} - 1)x'(\theta)$, where $\theta = \theta(s,h)$ is a midpoint lying in the interval $(\mathrm{e}^{-hs}s, s)$. We will prove that this expression converges uniformly to $-ay$, where $y \in C[0,1]$ is given by $y(s) = sx'(s)$ for $s > 0$ and $y(0) = 0$. To this end, we write

$$\begin{aligned}|sx'(\theta) - sx'(s)| &= |(s\theta^{-1} - 1)\theta x'(\theta) + \theta x'(\theta) - sx'(s)| \\ &\le |s\theta^{-1} - 1|\,|\theta x'(\theta)| + |\theta x'(\theta) - sx'(s)| \\ &\le (\mathrm{e}^{ah} - 1)\|y\| + |\theta x'(\theta) - sx'(s)|.\end{aligned}$$

Since $(0,1] \mapsto sx'(s)$ can be extended to a continuous function on $[0,1]$, we know that for any $\epsilon > 0$ there is a $\delta = \delta(\epsilon) > 0$ such that $|sx(s) - \widetilde{s}x'(\widetilde{s})| < \epsilon$ as long as the distance between $s, \widetilde{s} \in [0,1]$ is smaller than δ. Hence, the second summand in the right-most expression above does not exceed $\frac{\epsilon}{2}$ provided that h is so small that $1 - \mathrm{e}^{-ah} < \delta(\frac{\epsilon}{2})$; this is because $s - \theta(s,h) \le (1 - \mathrm{e}^{-ah})s \le (1 - \mathrm{e}^{-ah})$. The first summand in the rightmost expression does not depend on s or θ and can also be made $< \frac{\epsilon}{2}$ by taking h sufficiently small. This, when combined with $\lim_{h\to 0+} \frac{1-\mathrm{e}^{-ah}}{h} = a$, shows that for $x \in \mathcal{D}$,

$$\lim_{h\to 0+} h^{-1}[T(h)x - x] = -ay.$$

This concludes the proof that $\mathcal{D} \subset \mathcal{D}(A)$.

To summarize: the domain of the generator A of the semigroup of Exercise 15.14 (a) is composed of functions $x \in C[0,1]$ that are continuously differentiable at $s \in (0,1]$ and such that $\lim_{s\to 0} sx'(s) = 0$. Moreover, for such x's,

$$Ax(s) = -sx'(s), s \in (0,1] \qquad \text{and} \qquad Ax(0) = 0. \tag{16.9}$$

16.2.3 The Generator of the Semigroup of Exercise 15.16

In this case, it is most natural to seek a characterization of the domain $\mathcal{D}(A)$ of the generator of $\{T(t), t \ge 0\}$ in terms of the domain $\mathcal{D}(G)$ of the generator of $\{S(t), t \ge 0\}$. Hence, we note that

$$h^{-1}(T(h)x - x) - h^{-1}(S(h)x - x) = h^{-1}\mathrm{e}^{h} B \int_0^h \mathrm{e}^{-s} S(s)x \,\mathrm{d}s, \quad x \in \mathbb{X}, h > 0,$$

and, by (16.4), the right-hand side converges, as $h \to 0+$, to Bx, because B is bounded. This means that the limit $\lim_{h\to 0+} h^{-1}(T(h)x - x)$ exists whenever the $\lim_{h\to 0+} h^{-1}(S(h)x - x)$ does, and the two differ by Bx. Hence, x belongs to $\mathcal{D}(A)$ if and only if it belongs to $\mathcal{D}(G)$, and then we have

$$Ax = Gx + Bx. \tag{16.10}$$

It should perhaps be stressed here that the argument presented above never uses the requirements $B^2 = B$ and $S(t)B = B, t \ge 0$ imposed on B in the studied exercise. However, these requirements are crucially used in the proof that $\{T(t), t \geq 0\}$ is a semigroup. We close this subsection by noting that the second of them implies the following relation between G and B: for any $x \in \mathbb{X}$, Bx belongs to $\mathcal{D}(G)$ and $GBx = 0$. Indeed, under the assumption $S(t)B = B$, the map $t \mapsto S(t)Bx$ does not depend on t, and thus its derivative is 0. Conversely, if Bx belongs to $\mathcal{D}(G)$ and $GBx = 0$, formula (16.6) with $T(t)$ replaced by $S(t)$ and x replaced by Bx yields

$$S(t)Bx - Bx = \int_0^t S(s)GBx \, \mathrm{d}s = 0,$$

and this shows that $S(t)Bx = Bx, x \in \mathbb{X}$.

16.3 The Resolvent Equation

The examples presented above are used merely as illustrations, and are not meant to suggest that it is easy to characterize the generator of a given semigroup. Nevertheless, some kind of characterization of the generator is more often than not available when one cannot even dream of having a closed expression for the semigroup. In this context, it is vital to ask what properties of a given operator A guarantee that there is a semigroup for which A is the generator. We know from Section 16.1 that the minimum requirement is that A be closed and densely defined. This, however, will turn out to be merely a necessary condition.

The crucial property, discussed in the present section, involves the following integral, the Laplace transform of a semigroup's trajectory:

$$R_\lambda x := \int_0^\infty \mathrm{e}^{-\lambda t} T(t)x \, \mathrm{d}t. \tag{16.11}$$

The expression on the right-hand side is an improper integral, that is, it is understood as $\lim_{\tau \to \infty} \int_0^\tau \mathrm{e}^{-\lambda t} T(t)x \, \mathrm{d}t$. We claim that, given that $\{T(t), t \geq 0\}$ satisfies estimate (15.9), this improper integral exists for all $\lambda > \omega$. Indeed, for such a λ and any $\tau_2 > \tau_1$, we have

$$\left\| \int_{\tau_1}^{\tau_2} \mathrm{e}^{-\lambda t} T(t)x \, \mathrm{d}t \right\| \leq \int_{\tau_1}^{\tau_2} \|\mathrm{e}^{-\lambda t} T(t)x\| \, \mathrm{d}t \leq M\|x\| \int_{\tau_1}^{\tau_2} \mathrm{e}^{-(\lambda-\omega)t} \, \mathrm{d}t, x \in \mathbb{X}'.$$

Since $\int_0^\infty \mathrm{e}^{-(\lambda-\omega)t} \, \mathrm{d}t = \frac{1}{\lambda-\omega} < \infty$ it follows that for any sequence $(\tau_n)_{n\geq 1}$ of positive numbers diverging to infinity and any $x \in \mathbb{X}$, the sequence $(x_n)_{n\geq 1}$ of elements of $\mathbb{X}$ defined by

$$x_n = \int_0^{\tau_n} \mathrm{e}^{-\lambda t} T(t)x \, \mathrm{d}t$$

is fundamental in $\mathbb{X}$, and thus converges. Moreover, the same estimate shows that even though the sequences $(\tau_n)_{n\geq 1}$ might differ, the limit is always the same. This proves the claim: the integral on the right-hand side is well defined for all x and $\lambda > \omega$.

The family

$$R_\lambda, \lambda > \omega$$

is known as the *resolvent* of the semigroup. As a by-product of the reasoning presented above, we see that the R_λ's are bounded linear operators and

$$\|R_\lambda\| \leq \frac{M}{\lambda - \omega}. \tag{16.12}$$

The name *resolvent* comes from the fact that, for a $y \in \mathbb{X}$ and $\lambda > \omega$, the vector $x := R_\lambda y$ is the unique solution to the *resolvent equation* for the generator:

$$\lambda x - Ax = y. \tag{16.13}$$

Let's show that x indeed belongs to $\mathcal{D}(A)$ and solves the resolvent equation. To this end, we note that $S(t) := \mathrm{e}^{-\lambda t} T(t)$ defines a semigroup with generator $G = A - \lambda I_{\mathbb{X}}$ (in what follows we will write $A - \lambda$ instead of more proper $A - \lambda I_{\mathbb{X}}$); more precisely, the domain of G coincides with $\mathcal{D}(A)$, and for $z \in \mathcal{D}(A)$, we have $Gz = Az - \lambda z$. Formula (16.5) when applied to $S(t)$ shows that, for any natural n, $x_n := \int_0^n \mathrm{e}^{-\lambda t} T(t) y \, \mathrm{d}s$ belongs to $\mathcal{D}(A)$ and we have

$$(A - \lambda)x_n = \mathrm{e}^{-\lambda n} T(n)y - y.$$

As $n \to \infty$, the right-hand side here converges to $-y$ because $\|\mathrm{e}^{-\lambda n} T(n)\| \leq M\mathrm{e}^{-(\lambda-\omega)n}$. At the same time $\lim_{n\to\infty} x_n = R_\lambda y$. It follows that $\lim_{n\to\infty} Ax_n$ exists and equals $\lambda R_\lambda y - y$. Since A is closed, however, this means that $x = R_\lambda y$ belongs to $\mathcal{D}(A)$ and $Ax = \lambda x - y$, that is, that x solves the resolvent equation. In other words, R_λ is the right inverse to $\lambda - A$.

We proceed by showing that R_λ is also the left inverse for $\lambda - A$, that is, that there are no other solutions to the resolvent equation besides $R_\lambda y$. Suppose thus that x solves (16.13). Formula (16.6) with $T(t)$ replaced by $S(t)$ yields

$$\int_0^t \mathrm{e}^{-\lambda s} T(s)(Ax - \lambda x) \, \mathrm{d}s = \mathrm{e}^{-\lambda t} T(t)x - x.$$

Hence, letting $t \to \infty$ renders $R_\lambda(Ax - \lambda x) = -x$, and since by assumption $\lambda x - Ax = y$, we obtain $x = R_\lambda y$, as claimed.

> The resolvent equation for the generator of a semigroup satisfying (15.9) has precisely one solution for any $y \in \mathbb{X}$ and $\lambda > \omega$.

As we shall see later, the framed statement is a decisive property in examining candidates for a semigroup's generators.

16.4 Examples of Resolvents

Here are some examples of resolvents. For the semigroup of Exercise 15.14 (c), estimate (15.9) is satisfied with $M = 1$ and $\omega = 0$, and the resolvent is given by

$$R_\lambda(\eta_i)_{i\geq 1} = \int_0^\infty \mathrm{e}^{-\lambda t}\left(\mathrm{e}^{-ait}\eta_i\right)_{i\geq 1} \mathrm{d}t, \qquad \lambda > 0, (\eta_i)_{i\geq 1} \in \ell^2.$$

Since convergence in the norm of ℓ^2 implies convergence of coordinates, we obtain

$$R_\lambda(\eta_i)_{i\geq 1} = \left(\int_0^\infty \mathrm{e}^{-\lambda t}\mathrm{e}^{-ait}\eta_i \,\mathrm{d}t\right)_{i\geq 1} = \left(\frac{1}{\lambda + ai}\eta_i\right)_{i\geq 1}, \quad \lambda > 0, (\eta_i)_{i\geq 1} \in \ell^2.$$

This explicit formula confirms that $R_\lambda(\eta_i)_{i\geq 1}$ belongs to $\mathcal{D}(A)$ characterized in Section 16.2.1, because $\frac{i}{\lambda+ai} \leq a^{-1}$ for all integers i. Moreover,

$$\lambda R_\lambda(\eta_i)_{i\geq 1} - AR_\lambda(\eta_i)_{i\geq 1} = \lambda\left(\frac{1}{\lambda + ai}\eta_i\right)_{i\geq 1} + \left(\frac{ai}{\lambda + ai}\eta_i\right)_{i\geq 1} = (\eta_i)_{i\geq 1},$$

verifying that $R_\lambda(\eta_i)_{i\geq 1}$ solves the resolvent equation (with $y = (\eta_i)_{i\geq 1}$). Also, $\frac{\lambda}{\lambda+ai} \leq 1$ shows that

$$\|\lambda R_\lambda\| \leq 1;$$

this is a particular case of (16.12).

Similarly, in the case of the semigroup of Exercise 15.14 (a), we have

$$\begin{aligned} x(s) := (R_\lambda y)(s) &= \left(\int_0^\infty \mathrm{e}^{-\lambda t}T(t)y \,\mathrm{d}t\right)(s) = \int_0^\infty \mathrm{e}^{-\lambda t}T(t)y(s)\,\mathrm{d}t \\ &= \int_0^\infty \mathrm{e}^{-\lambda t}y(\mathrm{e}^{-at}s)\,\mathrm{d}t = a^{-1}s^{-\frac{\lambda}{a}}\int_0^s t^{\frac{\lambda}{a}-1}y(t)\,\mathrm{d}t, \qquad s > 0, \end{aligned}$$

and $x(0) = \lambda^{-1}y(0)$. A short calculation confirms that $\lim_{s\to 0+} x(s) = x(0)$ and so indeed $x \in C[0,1]$. Moreover, $x'(s)$ exists for $s \in (0,1]$ and equals $(as)^{-1}(y(s) - \lambda x(s))$. Furthermore, $\lim_{s\to 0+} sx'(s) = a^{-1}(y(0) - \lambda x(0)) = 0$, and thus x belongs to $\mathcal{D}(A)$, as characterized in Section 16.2.2. Finally, by

(16.9), $Ax(s) = -asx'(s) = \lambda x(s) - y(s), s > 0$ and $Ax(0) = \lambda x(0) - y(0)$, verifying that x solves the resolvent equation. A straightforward argument shows also that $\|\lambda x\| \le \|y\|$; this is another reflection of (16.12) with $\omega = 0$.

Finally, let's consider the semigroup of Exercise (15.16). Assuming without loss of generality that $\|S(t)\| \le M\mathrm{e}^{\omega t}$ for an $\omega > 1$, we obtain the following estimate for $\|T(t)\|$:

$$\begin{aligned}\|T(t)\| &\le M\mathrm{e}^{\omega t} + M\|B\|\mathrm{e}^{t}\int_0^t \mathrm{e}^{(\omega-1)s}\,\mathrm{d}s = M\mathrm{e}^{\omega t} + M\|B\|\mathrm{e}^{t}\frac{\mathrm{e}^{(\omega-1)t}-1}{\omega-1} \\ &\le M\mathrm{e}^{\omega t} + \frac{M\|B\|}{\omega-1}\mathrm{e}^{\omega t} =: \widetilde{M}\mathrm{e}^{\omega t}.\end{aligned}$$

Hence, the resolvent R_λ is defined at least for $\lambda > 1$, and we have[1]

$$\begin{aligned}R_\lambda y &= \int_0^\infty \mathrm{e}^{-\lambda t} S(t)y\,\mathrm{d}t + B\int_0^\infty \mathrm{e}^{-\lambda t}\int_0^t \mathrm{e}^{t-s}S(s)y\,\mathrm{d}s\,\mathrm{d}t \\ &= z + B\int_0^\infty\int_s^\infty \mathrm{e}^{-(\lambda-1)t}\mathrm{e}^{-s}S(s)y\,\mathrm{d}t\,\mathrm{d}s \\ &= z + (\lambda-1)^{-1}B\int_0^\infty \mathrm{e}^{-\lambda s}S(s)y\,\mathrm{d}s = z + (\lambda-1)^{-1}Bz, \qquad (16.14)\end{aligned}$$

where $z := \int_0^\infty \mathrm{e}^{-\lambda t}S(t)y\,\mathrm{d}t$ and in the second to last step we used Exercise 6.3.

Since, by definition, z belongs to the domain $\mathcal{D}(G)$ of the generator of the semigroup $\{S(t), t \ge 0\}$, and we know from Section 16.2.3 that Bz also belongs to $\mathcal{D}(G)$, this formula shows that so does $x := R_\lambda y$. Moreover, $\mathcal{D}(A)$ coincides with $\mathcal{D}(G)$ and, by (16.10),

$$\begin{aligned}\lambda x - Ax = \lambda x - (G+B)x &= \lambda x + \frac{\lambda}{\lambda-1}Bz - (G+B)z - \frac{1}{\lambda-1}(G+B)Bz \\ &= \lambda z - Gz\end{aligned}$$

because, by assumption, $B^2 = B$ and we know from Section 16.2.3 that $GBz = 0$. Since, by definition, $\lambda z - Gz = y$, x solves the resolvent equation for A, as expected.

16.5 The Generation Theorem

The sole requirement that the resolvent equation for a closed and densely defined operator A has a unique solution does not yet guarantee that A is

[1] In the second step we use a counterpart of the Fubini theorem for Riemann integrals of vector-valued functions; a full justification of this step requires the Hahn–Banach theorem and is given in the Appendix.

a generator. This assumption must be accompanied by the estimate on the norm of R_λ, as discussed below. The Hille–Yosida–Feller–Phillips–Miyadera generation theorem says that these two conditions together make A a generator.

We state the crucial estimate in the following lemma.

16.1 Lemma *Let $\{T(t), t \geq 0\}$ be a semigroup of class C_0 satisfying* (15.9). *Then*

$$\|R_\lambda^n\| \leq M(\lambda - \omega)^{-n}, \qquad n \geq 1, \lambda > 0.$$

Proof The following formula forms a basis for our argument:

$$R_\lambda^n y = \int_0^\infty \cdots \int_0^\infty \mathrm{e}^{-\lambda \sum_{i=1}^n t_i} T(\textstyle\sum_{i=1}^n t_i)\, \mathrm{d}t_n \ldots \mathrm{d}t_1, \quad \lambda > \omega, n \geq 1, y \in \mathbb{X}. \tag{16.15}$$

We will prove this relation by induction.

To this end, we note first that for $n = 1$ relation (16.15) reduces to the definition (16.11). Next, as a preparation for the induction step, we record that, for any $s \geq 0$, by continuity of $T(s)$ and (12.11),

$$\begin{aligned} T(s) \int_0^\infty \mathrm{e}^{-\lambda t} T(t) y \,\mathrm{d}t &= T(s) \lim_{n\to\infty} \int_0^n \mathrm{e}^{-\lambda t} T(t) y \,\mathrm{d}t = \lim_{n\to\infty} T(s) \int_0^n \mathrm{e}^{-\lambda t} T(t) y \,\mathrm{d}t \\ &= \lim_{n\to\infty} \int_0^n \mathrm{e}^{-\lambda t} T(t+s) y \,\mathrm{d}t = \int_0^\infty \mathrm{e}^{-\lambda t} T(t+s) y \,\mathrm{d}t. \end{aligned}$$

Hence, assuming that (16.15) holds for some $n \geq 1$, we can write

$$\begin{aligned} R_\lambda^{n+1} y &= R_\lambda^n R_\lambda y \\ &= \int_0^\infty \cdots \int_0^\infty \mathrm{e}^{-\lambda \sum_{i=1}^n t_i} T(\textstyle\sum_{i=1}^n t_i) \int_0^\infty \mathrm{e}^{-\lambda t_{n+1}} T(t_{n+1}) y \,\mathrm{d}t_{n+1} \,\mathrm{d}t_n \ldots \mathrm{d}t_1 \\ &= \int_0^\infty \cdots \int_0^\infty \mathrm{e}^{-\lambda \sum_{i=1}^{n+1} t_i} T(\textstyle\sum_{i=1}^{n+1} t_i) y \,\mathrm{d}t_{n+1} \ldots \mathrm{d}t_1. \end{aligned}$$

Since this establishes (16.15) for $n + 1$, the induction proof of (16.15) is completed.

The rest is now straightforward: by (16.15), (6.8) and (15.9),

$$\|R_\lambda^n y\| \leq M\|y\| \int_0^\infty \cdots \int_0^\infty \mathrm{e}^{-\lambda \sum_{i=1}^n t_i} \mathrm{e}^{\omega \sum_{i=1}^n t_i} \,\mathrm{d}t_n \ldots \mathrm{d}t_1 = \frac{M}{(\lambda - \omega)^n} \|y\|,$$

as desired. □

Notationally, R_λ in Lemma 16.1 refers to the Laplace transform (16.11), but we know from the previous section that $R_\lambda y$ in fact solves the resolvent equation (16.13). This realization leads to the following formulation of the generation theorem.

16.2 Theorem *(Hille–Yosida–Feller–Phillips–Miyadera) Let $A\colon \mathcal{D}(A) \to \mathbb{X}$ be a linear densely defined operator in a Banach space $\mathbb{X}$. A is the generator of a C_0-semigroup satisfying* (15.9) *if and only if the following conditions are satisfied.*

1. *For each $\lambda > \omega$ and $y \in \mathbb{X}$ there is precisely one $x \in \mathcal{D}(A)$ solving the resolvent equation*

$$\lambda x - Ax = y. \tag{16.16}$$

2. *The operator R_λ assigning the solution x of the resolvent equation to the y on its right-hand side is continuous and we have*

$$\|R_\lambda^n\| \leq M(\lambda - \omega)^{-n}, \qquad n \geq 1, \lambda > 0. \tag{16.17}$$

The necessity of these two conditions has already been established in Lemma 16.1. To wit, we know from the previous section that the resolvent equation for a semigroup's generator has a unique solution (as long as the λ featured in it is sufficiently large); this solution is in fact obtained by calculating the Laplace transform of the semigroup, and this forces the estimate (16.17), as shown in the lemma.

These conditions, however, are not only necessary but also sufficient, and this provides the following way of establishing that a given, densely defined operator is a generator. First of all, we study its resolvent equation for λ sufficiently large: we need to make sure that its solutions exist and are unique. If this is the case, we denote by R_λ the solution operator in the hope that it can be represented as the Laplace transform of the searched-for semigroup. This semigroup does indeed exist if estimates (16.17) hold.

We will prove the generation theorem in Section 16.7. Before completing the present section we note that in Theorem 16.2 there is no assumption that A is closed. However, it is in fact tacitly assumed in condition 2.: we know from Example 15.10 that continuity of R_λ implies closedness of A.

16.6 Contraction Semigroups and Dissipative Operators

In practice, for general M and ω, conditions of the Hille–Yosida–Feller–Phillips–Miyadera generation theorem are rather difficult to check. This is because, even if the resolvent equation can be solved explicitly (which is rare enough), the formula for R_λ^n is usually not available or too complicated to be useful. The task becomes simpler in the case of *contraction semigroups*, that is, for $M = 1$ and $\omega = 0$, when (16.17) reduces to just one inequality,

$$\|\lambda R_\lambda\| \le 1. \tag{16.18}$$

This is in fact the case to which the original papers of E. Hille and K. Yosida were devoted. W. Feller, R. Phillips and I. Miyadera extended Hille's and Yosida's result to general M and ω.

In this section we study in more detail such contraction semigroups: we will show that they are closely related to *dissipative operators*. By definition, an operator $A\colon \mathbb{X} \supset \mathcal{D}(A) \to \mathbb{X}$, where $\mathbb{X}$ is a Banach space, is termed dissipative if and only if

$$\|\lambda x - Ax\| \ge \lambda\|x\| \qquad \text{for all } \lambda > 0 \text{ and } x \in \mathcal{D}(A). \tag{16.19}$$

This notion nicely agrees with that introduced in Example 15.11: for a dissipative operator A in a Hilbert space and $\lambda > 0$, we have

$$\|\lambda x - Ax\|^2 = \lambda^2\|x\|^2 - 2\lambda(x, Ax) + \|Ax\|^2 \ge \lambda^2\|x\|^2, \qquad x \in \mathcal{D}(A),$$

and this means that A is also dissipative in the sense just introduced.

Here is the announced generation theorem for contraction semigroups.

16.3 Theorem *A densely defined operator A is the generator of a contraction semigroup in a Banach space $\mathbb{X}$ if and only if it is dissipative and there is a $\lambda_0 > 0$ such that for any $y \in \mathbb{X}$ there exists an $x \in \mathcal{D}(A)$ solving the resolvent equation.*

We stress the following crucial differences between Theorems 16.2 and 16.3. Most importantly: the task of proving that A is a generator using the latter is considerably simpler than in the case of the former. First of all, we need to study the resolvent equation for just one $\lambda_0 > 0$. Second, all we need to establish is the existence and not both the existence and uniqueness of its solutions. Finally, instead of attempting to prove difficult estimates (16.17), we need to simply check that A is dissipative.

Proof of Theorem 16.3 *(Necessity).* We need to show that A is dissipative, the rest being clear by Theorem 16.2. Thus, let $x \in \mathcal{D}(A)$ and $\lambda > 0$. Then, of course, x is a solution to the resolvent equation with right-hand side $y := \lambda x - Ax$. On the other hand, A being a contraction semigroup generator, this solution is unique, and given by (16.11) with $\|T(t)\| \le 1$. Therefore, $\|x\| \le \lambda^{-1}\|y\|$, and this means that $\lambda\|x\| \le \|\lambda x - Ax\|$. Since x is arbitrary, we have proved that A is dissipative.

(Sufficiency). We start by checking that solutions to the resolvent equation with $\lambda = \lambda_0$ are unique. Suppose that there are two solutions, say x' and x'', corresponding to one y. Then, for $x := x' - x''$ we have $\lambda_0 x - Ax = 0$. However, A being dissipative, we have $0 = \|\lambda_0 x - Ax\| \ge \lambda_0\|x\|$, and this

shows that $x = 0$, that is, that $x' = x''$. Hence, condition 1. in Theorem 16.2 is satisfied for $\lambda = \lambda_0$. Moreover, denoting by $R_{\lambda_0} y$ the solution to the resolvent equation (16.16) with $\lambda = \lambda_0$, we have $\lambda_0 \|R_{\lambda_0} y\| \leq \|y\|$; this is again because A is dissipative. We also conclude (see Example 15.10) that A is closed.

Next, let λ be such that $|\lambda - \lambda_0| < \lambda_0$ (i.e., $\lambda \in (0, 2\lambda_0)$). Then $|\lambda_0 - \lambda| \|R_{\lambda_0}\| < 1$, and thus the series

$$R_\lambda := \sum_{n=0}^{\infty} (\lambda_0 - \lambda)^n R_{\lambda_0}^{n+1}$$

converges in the operator norm of $\mathcal{L}(\mathbb{X})$. We will show that $R_\lambda y$ is a unique solution to the resolvent equation for $\lambda \in (0, \lambda_0)$. Indeed, for any integer $k \geq 1$ and $y \in \mathbb{X}$, $\sum_{n=0}^{k} (\lambda_0 - \lambda)^n R_{\lambda_0}^{n+1} y = R_{\lambda_0} (\sum_{n=0}^{k} (\lambda_0 - \lambda)^n R_{\lambda_0}^n y)$ belongs to $\mathcal{D}(A)$ and, writing $\lambda - A$ as $(\lambda - \lambda_0) + (\lambda_0 - A)$, we see that

$$\begin{aligned} (\lambda - A) \sum_{n=0}^{k} (\lambda_0 - \lambda)^n R_{\lambda_0}^{n+1} y &= (\lambda - \lambda_0) \sum_{n=0}^{k} (\lambda_0 - \lambda)^n R_{\lambda_0}^{n+1} y + \sum_{n=0}^{k} (\lambda_0 - \lambda)^n R_{\lambda_0}^{n} y \\ &= y - (\lambda_0 - \lambda)^{k+1} R_{\lambda_0}^{k+1} y. \end{aligned}$$

As $k \to \infty$, the right-hand side converges to y and, since A is closed, this shows that $(\lambda - A) R_\lambda y = y$, that is, that $R_\lambda y$ solves the resolvent equation. As before, we check that this solution is unique because A is dissipative.

Let's ponder on what we have proved. We have namely shown that the existence of solutions to the resolvent equation for a $\lambda_0 > 0$ implies the existence and uniqueness of solutions to this equation for all λ in the interval $(0, 2\lambda_0)$. It follows that the role of λ_0 can be played by any $\lambda \in (0, 2\lambda_0)$, and this in turn shows that the resolvent equation possesses unique solutions for all $\lambda \in (0, 4\lambda_0)$. An induction argument now establishes that condition 1. of Theorem 16.2 is satisfied for all $\lambda > 0$. Finally, dissipativity of A implies that $\|\lambda R_\lambda\| \leq 1$ and so condition 2. of Theorem 16.2 is satisfied with $M = 1$ and $\omega = 0$, completing the proof. □

16.4 Corollary *Any maximal dissipative operator in a Hilbert space is a contraction semigroup generator. In particular, any operator associated with a closed, non-negative bilinear form is a contraction semigroup generator.*

Proof We know from Example 15.11 that maximal dissipative operators are densely defined. Hence, the corollary is a particular case of (sufficiency part of) Theorem 16.3 with $\lambda_0 = 1$. □

Hence, operators from Examples 15.12 and 15.13 are generators of contraction semigroups, and so is the operator of Exercise 15.21.

16.5 Example Let $\mathbb{H} = L^2(\mathbb{R}^+)$ be the space of square integrable functions defined on the right half-axis $\mathbb{R}^+$. This is a Hilbert space with scalar product $(x, y) = \int_0^\infty x(\tau)y(\tau)\,\mathrm{d}\tau$. Also, let $\mathcal{D}(A)$ be composed of $x \in L^2(\mathbb{R}^+)$ such that y given by $y(\tau) = -\tau x(\tau), \tau \geq 0$ also belongs to $L^2(\mathbb{R}^+)$. We set $Ax = y$, and claim that A is maximal dissipative.

To this end, we note first of all that for $x \in \mathcal{D}(A)$, $(Ax, x) = -\int_0^\infty \tau[x(\tau)]^2\,\mathrm{d}\tau \leq 0$, proving that A is dissipative. Moreover, given $y \in L^2(\mathbb{R}^+)$ we introduce $x(\tau) = (1+\tau)^{-1}y(\tau), \tau \geq 0$. Then, x belongs to $\mathcal{D}(A)$ since

$$\int_0^\infty \left(\frac{\tau}{1+\tau}\right)^2 [x(\tau)]^2\,\mathrm{d}\tau \leq \int_0^\infty [x(\tau)]^2\,\mathrm{d}\tau < \infty,$$

and we have $x(\tau) - Ax(\tau) = y(\tau), \tau \geq 0$. This completes the proof of the claim. By Corollary 16.4 we conclude also that A is a contraction semigroup generator.

16.6 Example The operator A we encountered in Chapter 11 (see beginning of Section 11.4) when studying the heat equation, is also maximal dissipative. It should not be surprising that the formula (compare (11.15) and (11.16))

$$T(t)x = \sum_{n=1}^{\infty} \mathrm{e}^{-n^2 t} a_n \sin ns, \qquad t \geq 0, s \in [0, \pi], \tag{16.20}$$

where a_n's are Fourier coefficients of an $x \in L^2[0, \pi]$, defines the semigroup generated by A (see Exercise 16.8).[2]

The argument showing that A is dissipative is rather simple: for $x \in \mathcal{D}(A)$, we have $(Ax, x) = \int_0^\pi x''(s)x(s)\,\mathrm{d}s = -\int_0^\pi [x'(s)]^2\,\mathrm{d}s \leq 0$, where the second equality follows by integration by parts combined with the fact that $x(0) = x(\pi) = 0$ for $x \in \mathcal{D}(A)$. Maximality is not rocket science either: given $y \in L^2[0, \pi]$ we define $x(s) = C \sinh s - \int_0^s (s-u)y(u)\,\mathrm{d}u, s \in [0, \pi]$ with $C := \frac{\int_0^\pi (\pi - u)y(u)\,\mathrm{d}u}{\sinh \pi}$ to see that x belongs to $\mathcal{D}(A)$ and $x - Ax = y$.

For our next corollary to Theorems 16.2–16.3 we need the following notions.

Let $C(S)$ be the space of continuous functions on a compact topological space (see Chapter 7). An operator $A\colon C(S) \supset \mathcal{D}(A) \to C(S)$ is said to

[2] In this context, it is worth comparing the way the solution of the heat equation was understood in Chapter 11 with the summary of the connection between abstract Cauchy problems and semigroups given at the end of Section 16.1. Remarkably, the semigroup under study has the additional property expressed here in the fact that $T(t)x$ belongs to $\mathcal{D}(A)$ for $t > 0$ regardless of whether x does or does not. In fact, the semigroup belongs to an important class of holomorphic semigroups, which share this property (see e.g., [17, 18, 30] for a precise description of this class).

satisfy the *maximum principle* if and only if conditions $x \in \mathcal{D}(A)$ and $x(s_0) = \max_{s\in S} x(s)$ imply $Ax(s_0) \le 0$ (note that, for any $x \in C(S)$, there is an s_0 such that $x(s_0) = \max_{s\in S} x(s)$ precisely because S is a compact space).

Also, a strongly continuous semigroup $\{T(t), t \ge 0\}$ in $C(S)$ is said to be a conservative Feller semigroup if each $T(t)$ is a positive operator (i.e., $T(t)f \ge 0$ as long as $f \ge 0$) and $T(t)1_S = 1_S$ (here, $1_S(s) = 1$ for $s \in S$).

16.7 Corollary *A densely defined operator A in $C(S)$ is the generator of a conservative Feller semigroup if and only if*

(a) it satisfies the maximum principle,
(b) there is a $\lambda_0 > 0$ such that for all $y \in C(S)$, the resolvent equation for A with $\lambda = \lambda_0$ has a solution, and
(c) 1_S belongs to $\mathcal{D}(A)$ and $A1_S = 0$.

Proof *(Necessity)* For any $s \in S$, $x(s) \le x(s_0)$, that is, $x \le x(s_0)1_S$. Since $T(t)$'s are positive operators and $T(t)1_S = 1_S$, we have $T(t)x \le x(s_0)1_S$. On the other hand, if x belongs to $\mathcal{D}(A)$, the limit $\lim_{t\to 0} t^{-1}(T(t)x - x)$ exists in the sense of the norm of $C(S)$, and thus also pointwise. It follows that $Ax(s_0) = \lim_{t\to 0} t^{-1}(T(t)x(s_0) - x(s_0)) \le \lim_{t\to 0} t^{-1}(x(s_0) - x(s_0)) = 0$. This shows a).

Condition b) is clear by condition 1. in Theorem 16.2, since $T(t)$'s are contractions. For, inequality $\|x\| \le 1$ holds if and only if $-1_S \le x \le 1_S$, and the latter implies $-1_S \le T(t)x \le 1_S$ (i.e., $\|T(t)\| \le 1$), because $T(t)$'s are positive and $T(t)1_S = 1_S$.

To prove c), we note that difference quotients $t^{-1}(T(t)1_S - 1_S)$ are by assumption equal to 0.

(Sufficiency) Given $x \in \mathcal{D}(A)$ we can find an $s_0 \in S$ such that $x(s_0) = \max_{s\in S} x(s)$, and without loss of generality we can assume that $x(s_0) = \|x\|$ (if this equality does not hold, consider $-x$ instead). Then,

$$\begin{aligned}\|\lambda x - Ax\| &\ge \max_{s\in S} |\lambda x(s) - Ax(s)| \ge |\lambda x(s_0) - Ax(s_0)| \\ &= \lambda x(s_0) - Ax(s_0) \ge \lambda x(s_0) = \lambda\|x\|,\end{aligned}$$

the first equality following by the fact that both $x(s_0)$ and $-Ax(s_0)$ are non-negative.

This shows that A is dissipative, and we conclude by Theorem 16.3 that A generates a semigroup $\{T(t), t \ge 0\}$ of contractions. Moreover, by assumption c), the derivative of $t \mapsto T(t)1_S$ equals $T(t)A1_S = 0$, and thus this trajectory is constant, that is, $T(t)1_S = 1_S, t \ge 0$.

We are thus left with showing that $T(t)$'s are positive. Since these operators are limits of exponents of the Yosida approximation, it suffices to show that $R_\lambda y \ge 0$ for $\lambda > 0$ provided that $y \ge 0$. For a proof by contradiction, suppose that for x $(= R_\lambda y)$ there is an $s \in S$ such that $x(s) < 0$. Then there is also an s_0 such that $x(s_0) = \min_{s\in S} x(s) < 0$, and by the maximum principle we have $Ax(s_0) \ge 0$; this in turn implies $y(s_0) = \lambda x(s_0) - Ax(s_0) \le \lambda x(s_0) < 0$. This contradiction shows that $R_\lambda y$ is non-negative whenever y is. □

16.8 Example Let $C[0,1]$ be the space of continuous functions on the unit interval, and let the domain $\mathcal{D}(A)$ be composed of continuously differentiable functions x such that $x'(1) = a[x(0) - x(1)]$, where $a \ge 0$ is a given constant. Also, let $Ax = x'$. We will show that A is the generator of a conservative Feller semigroup.

Beginning with the maximum principle, we recall from real analysis that if the maximum of a function $x \in \mathcal{D}(A)$ is attained at an $s_0 \in (0,1)$, then necessarily $Ax(s_0) = x'(s_0) = 0$. Similarly, if $s_0 = 0$, then $x'(0) = \lim_{h\to 0+} h^{-1}(x(h) - x(0)) \le 0$. Finally, if $s_0 = 1$, then $Ax(1) = a(x(0) - x(1)) \le 0$; this establishes the maximum principle.

Since, clearly, 1_S belongs to $\mathcal{D}(A)$ and $A1_S = 0$ we are left with showing that for $\lambda > 0$ and a $y \in C[0,1]$ there is an $x \in \mathcal{D}(A)$ such that $\lambda x - x' = y$. We search for this x in the form $x(s) = C\mathrm{e}^{\lambda s} - \int_0^s \mathrm{e}^{\lambda(s-u)} y(u)\,\mathrm{d}u, s \in [0,1]$. It is easy to discover that this x belongs to $\mathcal{D}(A)$ for one and only one constant, namely

$$C = \frac{y(1) + (\lambda + a)\int_0^1 \mathrm{e}^{\lambda(1-u)} y(u)\,\mathrm{d}u}{(\lambda + a)\mathrm{e}^{\lambda} - 1}.$$

We can now appreciate afresh the fact that in Corollary 16.7 the assumption says 'there is a λ_0' and not 'for all $\lambda > 0$.' Were it not for this, we would have needed to cope with the unexpected difficulty: if $a < 1$ there is a $\lambda > 0$ such that the denominator in the definition of C vanishes. Fortunately, there are plenty of λ's for which the fraction is well defined, and so we are done by Corollary 16.7.

16.9 Remark The semigroup of the above example describes the following random process: a particle starting its motion at an $s \in [0,1)$ moves with constant speed to the right. After reaching $s = 1$ it stays there for an exponential time with parameter a, and then jumps to $s = 0$ to start all over again (forgetting the past completely).

16.7 Proof of the Generation Theorem

Since the necessity of conditions 1–2. in Theorem 16.2 has already been established, we only need to prove their sufficiency. The argument hinges on the properties of the *Yosida approximation*, which is composed of the following operators:

$$A_\lambda = \lambda^2 R_\lambda - \lambda, \qquad \lambda > \omega.$$

These are by definition bounded and thus we can think of their exponents $e^{tA_\lambda}, t \in \mathbb{R}$; the idea of the proof (due to Kosaku Yosida) is that as $\lambda \to \infty$, e^{tA_λ} converges to $T(t)$ for $t \geq 0$, where $\{T(t), t \geq 0\}$ is the searched-for semigroup generated by A.

We start by noting that an operator A satisfies conditions 1–2. of Theorem 16.2 for certain ω and M if and only if the operator $\widetilde{A}$, defined on $\mathcal{D}(A)$ by $\widetilde{A}x = Ax - \omega x$, satisfies these same conditions with $\omega = 0$. Moreover, if A is a generator then so is $\widetilde{A}$ and vice versa. In fact, the semigroup generated by A, say, $\{T(t), t \geq 0\}$, and that generated by $\widetilde{A}$, say, $\{\widetilde{T}(t), t \geq 0\}$, are related by $T(t) = e^{-\omega t}\widetilde{T}(t), t \geq 0$.

It follows that there is no loss in generality in assuming that we are dealing with $\omega = 0$; on the other hand, this assumption simplifies calculations considerably. Hence, from this point onward, we do assume $\omega = 0$, and gather the relevant properties of the Yosida approximation in the following proposition. The reader will notice again that the proof of point D) below breaks down if we do not assume that the space $\mathbb{X}$ in which we work is complete.

16.10 Lemma *Let conditions of Theorem 16.2 be satisfied with $\omega = 0$. Then,*

(A) $\|e^{tA_\lambda}\| \leq M, t \geq 0$,
(B) $\lim_{\lambda\to\infty} \lambda R_\lambda x = x, x \in \mathbb{X}$,
(C) $\lim_{\lambda\to\infty} A_\lambda x = Ax, x \in \mathcal{D}(A)$,
(D) *the limit $T(t)x := \lim_{\lambda\to\infty} e^{tA_\lambda}x$ exists for all $x \in \mathbb{X}$ and is uniform for t in any interval $[0, t_0] \subset \mathbb{R}^+$ (where $t_0 > 0$).*

Proof By assumption, for $t \geq 0$ and $\lambda > 0$,

$$\left\| \sum_{n=0}^{\infty} \frac{t^n}{n!}(\lambda^2 R_\lambda)^n \right\| \leq \sum_{n=0}^{\infty} \frac{t^n}{n!}\|(\lambda^2 R_\lambda)^n\| \leq M \sum_{n=0}^{\infty} \frac{\lambda^n t^n}{n!} = Me^{\lambda t}.$$

Therefore, by the second part of Exercise 13.12, $\|e^{tA_\lambda}\| = \|e^{-\lambda t}e^{t\lambda^2 R_\lambda}\| \leq M$, establishing A).

To prove B), similarly as in Section 14.6, it suffices to show convergence for x in a dense set, since $\|\lambda R_\lambda\| \leq M$. Now, $\mathcal{D}(A)$ is dense in $\mathbb{X}$, and for $x \in \mathcal{D}(A)$ we have $R_\lambda(\lambda x - Ax) = x$; it follows that $\|\lambda R_\lambda x - x\| = \|R_\lambda Ax\| \leq M\lambda^{-1}\|Ax\|$, and this converges to 0 as $\lambda \to \infty$.

B) in turn implies C): for $x \in \mathcal{D}(A)$ we have $A_\lambda x = \lambda(\lambda R_\lambda x - x) = \lambda R_\lambda Ax$ and this converges to Ax as $\lambda \to \infty$.

For the proof of D), we recall (see Lemma 13.6) that for any $\lambda > 0$, the function $s \mapsto \mathrm{e}^{sA_\lambda}$ is continuously differentiable with derivative $s \mapsto \mathrm{e}^{sA_\lambda}A_\lambda = A_\lambda \mathrm{e}^{sA_\lambda}$. Hence, for any $t > 0$ and $\lambda, \mu > 0$, so is the function $[0,t] \ni s \mapsto \mathrm{e}^{(t-s)A_\lambda}\mathrm{e}^{sA_\mu}$ and its derivative is $[0,t] \ni s \mapsto \mathrm{e}^{(t-s)A_\lambda}\mathrm{e}^{sA_\mu}(A_\mu - A_\lambda)$. It follows, by the second fundamental theorem of calculus, that

$$\mathrm{e}^{tA_\mu} - \mathrm{e}^{tA_\lambda} = \int_0^t \frac{\mathrm{d}}{\mathrm{d}s}(\mathrm{e}^{(t-s)A_\lambda}\mathrm{e}^{sA_\mu})\,\mathrm{d}s = \int_0^t \mathrm{e}^{(t-s)A_\lambda}\mathrm{e}^{sA_\mu}(A_\mu - A_\lambda)\,\mathrm{d}s,$$

and thus, for $x \in \mathbb{X}$,

$$\|\mathrm{e}^{tA_\mu}x - \mathrm{e}^{tA_\lambda}x\| \leq \left\| \int_0^t \mathrm{e}^{(t-s)A_\lambda}\mathrm{e}^{sA_\mu}(A_\mu - A_\lambda)x\,\mathrm{d}s \right\| \leq tM^2\|A_\mu x - A_\lambda x\|, \tag{16.21}$$

where we used A).

Now, by C), the right-hand side of (16.21) converges to 0, as $\lambda, \mu \to \infty$, provided that $x \in \mathcal{D}(A)$. Therefore, for any sequence $(\lambda_n)_{n\geq 1}$ converging to infinity, any $x \in \mathcal{D}(A)$ and any $t > 0$, $\left(\mathrm{e}^{tA_{\lambda_n}}x\right)_{n\geq 1}$ is a Cauchy sequence in $\mathbb{X}$ and thus converges. Moreover, the same estimate shows that the limit does not depend on the choice of $(\lambda_n)_{n\geq 1}$. We can thus define

$$T(t)x := \lim_{\lambda\to\infty} \mathrm{e}^{tA_\lambda}x, \qquad x \in \mathcal{D}(A), t \geq 0.$$

Then, letting μ tend to infinity in (16.21), we see that

$$\|T(t)x - \mathrm{e}^{tA_\lambda}x\| \leq tM^2\|Ax - A_\lambda x\|,$$

and this proves D) for $x \in \mathcal{D}(A)$. We actually have an explicit estimate of the rate of convergence in this case: for $t \in [0,t_0]$, $\|T(t)x - \mathrm{e}^{tA_\lambda}x\|$ does not exceed $M^2 t_0\|Ax - A_\lambda x\|, x \in \mathcal{D}(A)$.

The proof is completed as follows. By definition combined with A),

$$\|T(t)x\| \leq M\|x\|, \qquad t \geq 0, x \in \mathcal{D}(A). \tag{16.22}$$

Since $\mathcal{D}(A)$ is dense in $\mathbb{X}$, this allows us to extend each $T(t)$ to a bounded linear operator on the entire $\mathbb{X}$ by $T(t)x = \lim_{n\to\infty} T(t)x_n$, where $(x_n)_{n\geq 1}$ is a sequence of elements of $\mathcal{D}(A)$ converging to an $x \in \mathbb{X}$; inequality (16.22) makes it clear that the definition does not depend on the choice of $(x_n)_{n\geq 1}$, and we see that now (16.22) holds for all $x \in \mathbb{X}$.

Finally, given $x \in \mathbb{X}, t_0 > 0$ and $\epsilon > 0$, we choose an $\bar{x} \in \mathcal{D}(A)$ so that $\|x-\bar{x}\| < \frac{\epsilon}{4M}$. We know that for λ large enough, $\sup_{t\in[0,t_0]} \|T(t)\bar{x} - e^{tA_\lambda}\bar{x}\| \leq \frac{\epsilon}{2}$. But it follows that for such λ,

$$\begin{aligned}\sup_{t\in[0,t_0]} \|T(t)x - e^{tA_\lambda}x\| &\leq \sup_{t\in[0,t_0]} \|T(t)x - T(t)\bar{x}\| + \sup_{t\in[0,t_0]} \|T(t)\bar{x} - e^{tA_\lambda}\bar{x}\| \\ &\quad + \sup_{t\in[0,t_0]} \|e^{tA_\lambda}\bar{x} - e^{tA_\lambda}x\| \\ &\leq M\|x-\bar{x}\| + \tfrac{\epsilon}{2} + M\|x-\bar{x}\| < \epsilon.\end{aligned}$$

This shows point D), and completes the entire proof. □

Let's recapitulate our main findings. For each $x \in \mathbb{X}$, the function $[0,\infty) \ni t \mapsto T(t)x$ is continuous as a limit of continuous functions $t \mapsto e^{tA_\lambda}x$ that is uniform on any finite interval. Moreover, since $e^{tA_\lambda}e^{sA_\lambda} = e^{(t+s)A_\lambda}, s,t \geq 0$ (see Lemma 13.4), we have also $T(t)T(s) = T(t+s)$. In other words, $\{T(t), t \geq 0\}$ is a strongly continuous semigroup, and C) suggests that its generator is A.

To prove that it is really the case, given $x \in \mathcal{D}(A)$, we write (see Lemma 13.6)

$$e^{tA_\lambda}x - x = \int_0^t e^{sA_\lambda}A_\lambda x \,\mathrm{d}s, \qquad t \geq 0.$$

As $\lambda \to \infty$, the left-hand side here converges to $T(t)x - x$. As to the integrand on the right-hand side, it converges uniformly in $s \in [0,t]$ to $T(s)Ax$. This yields

$$T(t)x - x = \int_0^t T(s)Ax \,\mathrm{d}s, \qquad t \geq 0, x \in \mathcal{D}(A),$$

and then (6.5) shows that x is in the domain, say, $\mathcal{D}(\widetilde{A})$, of the infinitesimal generator $\widetilde{A}$ of $\{T(t), t \geq 0\}$ and that $\widetilde{A}x = Ax$; in other words, $\widetilde{A}$ extends A.

We are left with arguing that $\widetilde{A}$ is not a proper extension of A. For a proof by contradiction, suppose there is an $x \in \mathcal{D}(\widetilde{A})$ that does not belong to $\mathcal{D}(A)$. By condition 1. of Theorem 16.2 there is an $\bar{x} \in \mathcal{D}(A)$ solving the resolvent equation for A with $y := \lambda x - \widetilde{A}x$:

$$\lambda\bar{x} - A\bar{x} = y.$$

Since $\widetilde{A}$ extends A we have found two distinct solutions to the same resolvent equation: $x \in \mathcal{D}(A)$ and $\bar{x} \notin \mathcal{D}(A)$. This is, however, impossible, since $\widetilde{A}$ is a generator. This contradiction shows that $A = \widetilde{A}$ and thus completes the entire proof of the generation theorem.

16.8 Generation Theorems for Positive Semigroups

As already mentioned, it is next to impossible to check the main estimate in the Hille–Yosida–Feller–Phillips–Miyadera generation theorem for general M and ω. Amazingly, there are rather simple criteria that allow checking that a *resolvent positive operator* is a generator. In this section, following [4, 5, 8, 9], we present three results of this type.

These theorems are best discussed in the set-up of the general theory of Banach lattices, but here, for simplicity of exposition, we restrict ourselves to analysis in the spaces (serving as basic examples of Banach lattices) of continuous and integrable functions, respectively, where abstract definitions have a direct intuitive meaning.

An operator, say, A, in one of these spaces is said to be *resolvent positive* if there is a real ω such that (a) the resolvent equation

$$\lambda x - Ax = y$$

has a unique solution $x \in \mathcal{D}(A)$ for every y and $\lambda > \omega$, and (b) the solution x is non-negative whenever y is.

As before, we write $x = R_\lambda y$; it can be checked (see e.g., Thm. 2.65 in [8]) that then $R_\lambda, \lambda > \omega$ are bounded linear operators. By assumption they are also *positive* in that $R_\lambda y \geq 0$ whenever $y \geq 0$.

16.11 Theorem *Let S be a compact topological space. Any densely defined, resolvent positive operator A in $C(S)$ (the space of continuous functions on S) generates a semigroup, and the semigroup it generates is composed of positive operators.*

Proof

(I) By Exercise 15.13, the Hilbert equation is satisfied: $(\lambda - \mu)R_\lambda R_\mu = R_\mu - R_\lambda$ for $\lambda, \mu > \omega$, and thus we can write

$$\begin{aligned} R_\mu &= R_\lambda + (\lambda - \mu)R_\lambda R_\mu \\ &= R_\lambda + (\lambda - \mu)R_\lambda(R_\lambda + (\lambda - \mu)R_\lambda R_\mu) \\ &= R_\lambda + (\lambda - \mu)R_\lambda^2 + (\lambda - \mu)^2 R_\lambda^2 R_\mu, \qquad \lambda > \mu > \omega. \end{aligned}$$

An induction argument now shows that for any $n \geq 1$ and $\lambda > \mu > \omega$,

$$R_\mu = R_\lambda + (\lambda - \mu)R_\lambda^2 + \cdots + (\lambda - \mu)^{n-1}R_\lambda^n + (\lambda - \mu)^n R_\lambda^n R_\mu. \quad (16.23)$$

(II) For any $x \in C(S)$, $|x|$ defined by $|x|(s) := |x(s)|, s \in S$ belongs to $C(S)$ and, of course, $-|x|(s) \leq x(s) \leq |x|(s)$. Since R_λ is a positive operator, so are R_λ^n for all $n \geq 1$, and thus the above inequality implies, by linearity, that$-R_\lambda^n|x| \leq R_\lambda^n x \leq R_\lambda^n|x|$, that is, that

$$|R_\lambda^n x| \le R_\lambda^n |x|.$$

(III) Since A is densely defined, we can find an element x_0 of $\mathcal{D}(A)$ within $\frac{1}{2}$ distance from the function that equals 2 everywhere on S. In particular, $x_0(s) > 1$ for all $s \in S$. We fix $\mu > \omega$ and let $y_0 = \mu x_0 - Ax_0$. Then, by (II), for any $x \in C(S)$ with $\|x\| \le 1$ and $\lambda > \omega$,

$$\begin{aligned}(\lambda - \mu)^n |R_\lambda^n x| &\le (\lambda - \mu)^n R_\lambda^n |x| \le (\lambda - \mu)^n R_\lambda^n x_0 = (\lambda - \mu)^n R_\lambda^n R_\mu y_0 \\ &\le (\lambda - \mu)^n R_\lambda^n R_\mu |y_0| \le R_\mu |y_0|. \end{aligned} \tag{16.24}$$

The last inequality in the calculation above is a result of (16.23), because all the functions $R_\lambda |y_0|, (\lambda - \mu) R_\lambda^2 |y_0|, \dots, (\lambda - \mu)^{n-1} R_\lambda^n |y_0|$ are non-negative.

(IV) For $M := \|R_\mu |y_0|\|$, (16.24) renders $\|(\lambda - \mu)^n R_\lambda^n x\| \le M$ for $\|x\| \le 1$, that is, $\|(\lambda - \mu)^n R_\lambda^n\| \le M$. This proves, by the Hille–Yosida–Feller–Phillips–Miyadera theorem, that A is a generator. Moreover, since the semigroup $\{T(t), t \ge 0\}$ this operator generates is obtained as the limit

$$T(t) = \lim_{\lambda \to \infty} \mathrm{e}^{-\lambda t} \mathrm{e}^{t \lambda^2 R_\lambda} = \lim_{\lambda \to \infty} \mathrm{e}^{-\lambda t} \sum_{n=0}^{\infty} \frac{t^n \lambda^{2n}}{n!} R_\lambda^n$$

and R_λ^n are positive operators, so are $T(t)$. □

The next theorem concerns the space of integrable functions. More specifically, let $(S, \mathcal{F}, m)$ be a space with measure; readers that are not familiar with the general theory of integration may think of $S = \mathbb{R}^k$ and the k-dimensional Lebesgue measure, or S being a subset of $\mathbb{R}^k$ with the measure inherited from $\mathbb{R}^k$. By L^1 we denote the Banach space of (equivalence classes of) integrable functions with norm

$$\|x\| := \int_S |x| \, \mathrm{d}m.$$

16.12 Theorem *Let a densely defined A in L^1 be resolvent positive, and assume that for some $\mu > \omega$ there is a $c > 0$ such that*

$$\int_S R_\mu x \, \mathrm{d}m \ge c \int_S x \, \mathrm{d}m \tag{16.25}$$

for all non-negative $x \in L^1$. Then A generates a semigroup of positive operators.

Proof Since, by the Hilbert equation, $R_\lambda R_\mu = R_\mu R_\lambda$, and (16.23) is still in force, we can argue as before that for any $\lambda > \mu$ and $x \ge 0$,

$$\int_S (\lambda - \mu)^n R_\mu R_\lambda^n x \, \mathrm{d}m = \int_S (\lambda - \mu)^n R_\lambda^n R_\mu x \, \mathrm{d}m \le \int_S R_\mu x \, \mathrm{d}m.$$

Furthermore, $\int_S (\lambda-\mu)^n R_\lambda^n x \,\mathrm{d}m \le c^{-1}\int_S (\lambda-\mu)^n R_\mu R_\lambda^n x \,\mathrm{d}m$, by assumption. Therefore,

$$\int_S (\lambda-\mu)^n R_\lambda^n x \,\mathrm{d}m \le c^{-1}\int_S R_\mu x \,\mathrm{d}m, \qquad x \ge 0, \lambda > \omega. \tag{16.26}$$

It suffices now to show that this estimate implies

$$\|(\lambda-\mu)^n R_\lambda^n\| \le c^{-1}\|R_\mu\|.$$

(Positivity of the semigroup generated by A is proved as in the previous theorem.) To this end, we fix $\lambda > \mu$ and let, $P_n := (\lambda-\mu)^n R_\lambda^n$, for notational simplicity; in particular, (16.26) can now be written in brief as $\|P_n x\| \le c^{-1}\|R_\mu x\|, x \ge 0$. Also, for $x \in L^1$ we introduce x^+ and x^- defined by $x^+(s) = \max(x(s),0)$ and $x^-(s) = \max(-x(s),0), s \in S$, so that $x = x^+ - x^-$ and $\|x\| = \|x^+\| + \|x^-\|$. By definition, $x^+ \ge x$ and $x^- \ge -x$, and hence $P_n x^+ \ge P_n x$ and $P_n x^- \ge P_n(-x)$. Thus

$$0 \le (P_n x)^+ \le P_n x^+ \qquad \text{and} \qquad 0 \le (P_n x)^- \le P_n x^-.$$

This justifies the following calculation

$$\begin{aligned}\|P_n x\| &= \|(P_n x)^+\| + \|(P_n x)^-\| \le \|P_n x^+\| + \|P_n x^-\| \le c^{-1}(\|R_\mu x^+\| + \|R_\mu x^-\|)\\ &\le c^{-1}\|R_\mu\|(\|x^+\| + \|x^-\|) = c^{-1}\|R_\mu\|\|x\|.\end{aligned}$$

Since this holds for all n and $x \in L^1$, we are done. □

If condition (16.25) is satisfied for all $\mu > 0$ with $c = c(\mu) = \mu^{-1}$ and inequality replaced by equality, the operators $\mu R_\mu, \mu > 0$ are said to be *Markov operators*. Markov operators play an important role in the theory of stochastic processes and in modeling natural phenomena [12, 25, 36]. As a corollary to Theorem 16.12 we see that a resolvent positive operator with the property that all $\mu R_\mu, \mu > 0$ are Markov operators is a positive semigroup generator. Straightforward reasoning based on the Yosida approximation shows that the semigroup such an A generates is also composed of Markov operators.

Here is the last criterion for a resolvent positive operator to be a generator.

16.13 Theorem *Let A be a resolvent positive, densely defined operator in L^1, and suppose that for every $y \ge 0$ there is an $x \in \mathcal{D}(A)$ such that $x \ge y$. Then A is a positive semigroup generator.*

Proof Let $\lambda > \mu > \omega$ and $y \geq 0$ in L^1 be fixed. There is an $x \in \mathcal{D}(A)$ such that $y \leq x$, and we consider $y_1 := \mu x - Ax$. Then

$$(\lambda - \mu)^n R_\lambda^n y \leq (\lambda - \mu)^n R_\lambda^n x = (\lambda - \mu)^n R_\lambda^n R_\mu y_1 \leq (\lambda - \mu)^n R_\lambda^n R_\mu |y_1| \\ \leq R_\mu |y_1|,$$

with the last step following again by (16.23).

This shows that $\sup_{\lambda > \mu, n \geq 1} \|(\lambda - \mu)^n R_\lambda^n y\| < \infty$ whenever $y \geq 0$. Since any $y \in L^1$ can be written as $y = y^+ - y^-$ with $y^+, y^- \geq 0$, as in the proof of Theorem 16.12, we see that this property is shared by all $y \in L^1$. The Banach–Steinhaus theorem now shows that there is an M such that $\sup_{\lambda > \mu, n \geq 1} \|(\lambda - \mu)^n R_\lambda^n\| \leq M$, and we conclude that A is a generator. Positivity of the semigroup A generates is again deduced from the Yosida approximation. □

16.9 Exercises

Exercise 16.1. Prove that the inequality $\|T(t)\| \leq Me^{\omega t}, t > 0$ (excluding $t = 0$, compare (15.9)) cannot hold for a strongly continuous semigroup unless $M \geq 1$. **Hint:** Write a $t > 0$ as $t = \frac{t}{n} + \frac{t}{n} + \cdots + \frac{t}{n}$ and use the semigroup property to see that $M < 1$ forces $T(t) = 0$.

Exercise 16.2. Let $\{T(t), t \geq 0\}$ be a strongly continuous semigroup of operators with generator A.

(a) Let a be a positive real number. Show that $\{T_a(t), t \geq 0\}$ where $T_a(t) = T(at), t \geq 0$ is also a strongly continuous semigroup. Moreover, its generator, say, B, is characterized as follows: its domain coincides with the domain of A, and $Bx = aAx$ for x in this domain.

(b) Let a be a real number. Show that $\{T_a(t), t \geq 0\}$ where $T_a(t) = \mathrm{e}^{at} T(t), t \geq 0$ is also a strongly continuous semigroup. Moreover, its generator, say B, is characterized as follows: its domain coincides with the domain of A, and $Bx = ax + Ax$ for x in this domain.

Exercise 16.3. Prove that the domain of the generator of the semigroup introduced in Exercise 15.14 (b) is composed of continuously differentiable $x \in C[0, \infty]$ such that $x' \in C[0, \infty]$. Moreover $Ax = ax'$.

Exercise 16.4. Characterize the generator of the semigroup $\{T(t), t \geq 0\}$ of Exercise 15.17. Show also that condition $BS(t) = B, t \geq 0$ is equivalent to the following relation between B and the generator G of $\{S(t), t \geq 0\}$: for

$x \in \mathcal{D}(G)$, we have $BGx = 0$. **Hint:** For $x \in \mathcal{D}(G)$, we have $S(t)x - x = \int_0^t GS(s)x\,\mathrm{d}s$.

Exercise 16.5. Find an explicit formula for the resolvent of the operator of Example 16.5, and for the related Yosida approximation. What is the formula for the semigroup generated by A?

Exercise 16.6. Let $\omega \in \mathbb{R}$ and let $\mathbb{X}$ be a Banach space. A family $R_\lambda, \lambda > \omega$ of bounded linear operators in $\mathbb{X}$ is termed *pseudoresolvent* if it satisfies the Hilbert equation:

$$R_\lambda - R_\mu = (\mu - \lambda)R_\lambda R_\mu, \qquad \lambda, \mu > \omega.$$

It is said to be tempered at infinity if, for each $x \in \mathbb{X}$, $\lim_{\lambda\to\infty} R_\lambda x = 0$. Check that for any pseudoresolvent tempered at infinity,

$$\lim_{\lambda\to\infty} \lambda R_\lambda R_\mu x = R_\mu x, \qquad \mu > \omega, x \in \mathbb{X}.$$

Exercise 16.7. Let A be the generator of a bounded strongly continuous semigroup, and let $A_\lambda, \lambda > 0$ be its Yosida approximation. Check to see that

$$(A_\lambda - A)R_\mu = (\lambda R_\lambda - I)(\mu R_\mu - I), \qquad \lambda, \mu > 0.$$

Exercise 16.8. Prove that formula (16.20) defines a semigroup of contraction operators with generator A defined in Section 11.4. Is the operator B of Section 11.5 maximal dissipative also? What is the formula for the semigroup it generates?

Exercise 16.9. Let $\{T(t), t \geq 0\}$ be a strongly continuous semigroup in a Banach space $\mathbb{X}$, and let $\{S(t), t \geq 0\}$ be a strongly continuous semigroup in a Banach space $\mathbb{Y}$. These semigroups are termed *isomorphic* (or *similar*) if there is an isomorphism $I : \mathbb{X} \to \mathbb{Y}$ (i.e., I is surjective and injective) such that

$$S(t) = IT(t)I^{-1}, \qquad t \geq 0.$$

1. Characterize the generator of the semigroup $\{S(t), t \geq 0\}$ in terms of the generator of $\{T(t), t \geq 0\}$, assuming that they are isomorphic.
2. Prove that the semigroups introduced in Exercise 15.14 (a) and (b) are isomorphic.
3. Prove that the semigroups generated by the operators A and B discussed in Chapter 11 are isomorphic. **Hint:** See Exercise 12.9.
4. Let $\mathbb{X}$ be the space $C[0, 1]$ of continuous functions on the unit interval, and let $\mathcal{D}(A)$ be composed of continuously differentiable functions x such that $x'(0) = a[x(1) - x(0)]$, where $a \geq 0$ is a given constant. Also, let $Ax = -x'$. Show that A is the generator of a conservative Feller semigroup that

is isomorphic to the semigroup of Example 16.8. How would you describe the related stochastic process?

Exercise 16.10. Let $L^2[0,1]$ be the Hilbert space of square integrable functions $x\colon [0,1] \to \mathbb{R}$ with scalar product $(x,y) = \int_0^1 x(s)y(s)\,\mathrm{d}s$. Let $\mathcal{D}(A)$ be composed of functions $x \in L^2[0,1]$ that are continuously differentiable and such that $x'(s) = x'(0) + \int_0^s y(u)\,\mathrm{d}u$ for some $y \in L^2[0,1]$; moreover $x'(0) = x'(1) = 0$ (these are *Neumann boundary conditions* describing *reflecting Brownian motion*). For such x, we define $Ax = y(= x'')$. Prove that this operator is maximal dissipative and thus generates a contraction semigroup in $L^2[0,1]$.

Exercise 16.11. ▲ Let $C[0,2]$ be the space of continuous functions on the closed interval $[0,2]$, and let $\mathcal{D}(A)$ be composed of $x \in C[0,2]$ that are twice continuously differentiable and such that $x'(2) = 0$ whereas $x''(0) = a(x(1) - x(0))$ for a given $a \geq 0$; we let $Ax = x''$. Show that A is a conservative Feller generator.[3]

Exercise 16.12. ▲ Let $A_\lambda, \lambda > 0$ be the Yosida approximation of the generator A of a strongly continuous semigroup satisfying (15.9) with $\omega = 0$. Check that

$$(\mu - A_\lambda)^{-1} = \tfrac{1}{\lambda+\mu} + (\tfrac{\lambda}{\lambda+\mu})^2 R_{\frac{\lambda\mu}{\lambda+\mu}}, \qquad \mu > 0.$$

Hint: Use the resolvent equation for A.

☞ CHAPTER SUMMARY

The chapter is a gentle introduction to the theory of strongly continuous semigroups of operators. We present the notion of the generator, discuss the generator's basic properties and study a number of examples. We learn that the way to discover whether a given operator is a semigroup generator is by examining the resolvent equation, and are thus naturally led to the Hille–Yosida–Feller–Phillips–Miyadera theorem that characterizes generators in terms of resolvents. Two valuable consequences, the generation theorems for maximal dissipative operators in Hilbert space and operators satisfying the positive-maximum principle in the space of continuous function are also

[3] The related process is a Brownian motion that at $s = 2$ is reflected and at $s = 0$ is temporarily captured. More precisely, after reaching $s = 0$ the process stays there for an exponential time with parameter $a = 0$ and then starts all over again from $s = 1$, forgetting the past completely.

explained. This material is supplemented with three theorems on the generation of positive semigroups. The reader of this book, however, will undoubtedly have noticed that the whole theory would have failed were it not for the fact that we are working in Banach spaces; without the assumption of completeness, we could not be sure that the Yosida approximation (of Section 16.7) converges, and the entire reasoning would have collapsed.

Appendix: Two Consequences of the Hahn–Banach Theorem

Let $\mathbb{X}$ be a linear space. A linear map $F\colon \mathbb{X} \to \mathbb{R}$ is said to be a functional. If $\mathbb{X}$ is normed and F is bounded, F is referred to as continuous functional. The Hahn–Banach theorem, one of the fundamental results of functional analysis, allows us to extend some functionals that are defined merely on a subspace of $\mathbb{X}$ to the entire $\mathbb{X}$, and thus reveals that there is an abundant number of continuous linear functionals in any Banach space. It is one of the consequences of this theorem that for any distinct $x_1, x_2 \in \mathbb{X}$ there is a continuous linear functional F such that

$$Fx_1 \neq Fx_2. \tag{A.1}$$

Put otherwise, if $Fx_1 = Fx_2$ for all bounded, linear functionals, then $x_1 = x_2$.

In this appendix, we use this fact to fill the gaps in the proofs of two results presented in the book: one of Section 6.3 and one of Section 16.5.

A.1 Vector-Valued Function with Zero Derivative Is Constant

Suppose $[a,b] \ni t \mapsto x(t) \in \mathbb{X}$ is continuously differentiable with zero derivative. Let $F\colon \mathbb{X} \to \mathbb{R}$ be a continuous linear functional. The real-valued function $t \mapsto F(x(t))$ is then continuously differentiable with $\frac{\mathrm{d}}{\mathrm{d}t}F(x(t)) = F(x'(t)) = 0$. Thus, invoking the well-known result for real-valued functions, we see that $F(x(t_1)) = F(x(t_2))$ for any two points $t_1, t_2 \in [a,b]$. Since F is arbitrary, we must have $x(t_1) = x(t_2)$ and this shows the claim.

A.2 A Gap in the Calculation of Section 16.3

Here, we justify the calculation of (16.14), by proving that

$$\int_0^\infty \mathrm{e}^{-\lambda t} \int_0^t \mathrm{e}^{t-s} S(s)y \,\mathrm{d}s\,\mathrm{d}t = \int_0^\infty \int_s^\infty \mathrm{e}^{-(\lambda-1)t}\mathrm{e}^{-s} S(s)y \,\mathrm{d}t\,\mathrm{d}s. \tag{A.2}$$

To this end, we take a bounded linear functional F and apply it to the left-hand side. By Exercise 12.7 and Fubini's theorem we obtain

$$\int_0^\infty \mathrm{e}^{-\lambda t} \int_0^t \mathrm{e}^{t-s} F[S(s)y]\,\mathrm{d}s\,\mathrm{d}t = \int_0^\infty \int_s^\infty \mathrm{e}^{-(\lambda-1)t}\mathrm{e}^{-s} F[S(s)y]\,\mathrm{d}t\,\mathrm{d}s.$$

Next, we note, again by Exercise 12.7, that applying F to the right-hand side of (A.2) yields the same value. Since F is arbitrary, this proves (A.2).

References

[1] Y. A. Abramovich and C. D. Aliprantis. *An Invitation to Operator Theory*, volume 50 of Graduate Studies in Mathematics. American Mathematical Society, 2002. (Cited on p. 49.)

[2] R. A. Adams and J. J. F. Fournier. *Sobolev Spaces*, volume 140 of Pure and Applied Mathematics (Amsterdam). Elsevier/Academic Press, second edition, 2003. (Cited on pp. 116 and 117.)

[3] B. Åkerberg. Classroom notes: A proof of the arithmetic-geometric mean inequality. *Amer. Math. Monthly*, 70(9):997–998, 1963. (Cited on p. 2.)

[4] W. Arendt. Resolvent positive operators. *Proc. Lond. Math. Soc. (3)*, 54:321–349, 1987. (Cited on p. 231.)

[5] W. Arendt, C. J. K. Batty, M. Hieber, and F. Neubrander. *Vector-Valued Laplace Transforms and Cauchy Problems*. Birkhäuser, 2001. (Cited on p. 231.)

[6] S. Banach. Über die Baire'sche Kategorie gewisser Funktionenmengen. *Studia Mathematica*, 3:174–179, 1931. (Cited on p. 187.)

[7] S. Banach and H. Steinhaus. Sur le principe de la condensation de singularités. *Fundamenta Mathematica*, 9:50–61, 1927. (Cited on p. 185.)

[8] J. Banasiak and L. Arlotti. *Perturbations of Positive Semigroups with Applications*. Springer, 2006. (Cited on p. 231.)

[9] C. J. K. Batty and D. W. Robinson. Positive one-parameter semigroups on ordered Banach spaces. *Acta Appl. Math.*, 2:221–296, 1984. (Cited on p. 231.)

[10] A. Bielecki. Une remarque sur la méthode de Banach–Cacciopoli–Tikhonov. *Bull. Polish Acad. Sci.*, 4:261–268, 1956. (Cited on pp. 20 and 29.)

[11] A. Bobrowski. *Functional Analysis for Probability and Stochastic Processes. An Introduction*. Cambridge University Press, 2005. (Cited on pp. 79, 100, 166 and 180.)

[12] A. Bobrowski. *Generators of Markov Chains. From a Walk in the Interior to a Dance on the Boundary*. Cambridge University Press, 2021. (Cited on pp. 180 and 233.)

[13] N. L. Carothers. *A Short Course on Banach Space Theory*, volume 64 of London Mathematical Society Student Texts. Cambridge University Press, 2004. (Cited on pp. 166 and 185.)

[14] J. B. Conway. *A Course in Functional Analysis*, volume 96 of Graduate Texts in Mathematics. Springer-Verlag, second edition, 1990. (Cited on p. 77.)

[15] R. Durrett. *Probability: Theory and Examples*. Cambridge Series in Statistical and Probabilistic Mathematics. Cambridge University Press, fourth edition, 2010. (Cited on p. 180.)

[16] R. E. Edwards. *Functional Analysis. Theory and Applications*. Dover Publications, 1995. (Cited on pp. 20 and 29.)

[17] K.-J. Engel and R. Nagel. *One-Parameter Semigroups for Linear Evolution Equations*. Springer, 2000. (Cited on p. 225.)

[18] K.-J. Engel and R. Nagel. *A Short Course on Operator Semigroups*. Springer, 2006. (Cited on p. 225.)

[19] L. C. Evans. *Partial Differential Equations*, volume 19 of Graduate Studies in Mathematics. American Mathematical Society, second edition, 2010. (Cited on p. 115.)

[20] W. Feller. *An Introduction to Probability Theory and Its Applications*, volume 1. Wiley, 1950. Third edition, 1970. (Cited on pp. 153, 158 and 170.)

[21] J. Górnicki. *Okruchy matematyki*. Wydawnictwo Naukowe PWN, 1995. (Cited on pp. 18 and 29.)

[22] E. Hille and R. S. Phillips. *Functional Analysis and Semi-Groups*. American Mathematical Society Colloquia Publication 31. American Mathematical Society. 1957. (Cited on p. 65.)

[23] O. Kallenberg. *Foundations of Modern Probability*. Springer, 1997. (Cited on p. 100.)

[24] T. Kato. *Perturbation Theory for Linear Operators*. Classics in Mathematics Series. Springer, 1995, reprint of the 1980 edition. (Cited on p. 204.)

[25] A. Lasota and M. C. Mackey. *Chaos, Fractals, and Noise: Stochastic Aspects of Dynamics*. Springer, 1994. (Cited on p. 233.)

[26] A. Lasota and J. Myjak. Generic properties of stochastic semigroups. *Bull. Polish Acad. Sci. Math.*, 40(4):283–292, 1992. (Cited on p. 187.)

[27] N. N. Lebedev. *Special Functions and Their Applications*. Dover 1972. Revised edition, translated from the Russian and edited by Richard A. Silverman, unabridged and corrected republication. (Cited on p. 111.)

[28] L. Maligranda. Why Hölder's inequality should be called Rogers' inequality. *Math. Inequal. Appl.*, 1(1):69–83, 1998. (Cited on p. 2.)

[29] S. Mazurkiewicz. Sur les fonctions non dérivables. *Studia Mathematica*, 3:92–94, 1931. (Cited on p. 187.)

[30] A. Pazy. *Semigroups of Linear Operators and Applications to Partial Differential Equations*. Springer, 1983. (Cited on p. 225.)

[31] A. Pietsch. *History of Banach Spaces and Linear Operators*. Birkhäuser Boston 2007. (Cited on pp. 166 and 185.)

[32] M. A. Pinsky. *Partial Differential Equations and Boundary-Value Problems with Applications*, volume 15 of Pure and Applied Undergraduate Texts. American Mathematical Society, 2011. Reprint of the third (1998) edition. (Cited on p. 115.)

[33] M. Reed and B. Simon. *Functional Analysis*, volume 1 of Methods of Modern Mathematical Physics. Academic Press [Harcourt Brace Jovanovich], second edition, 1980. (Cited on p. 166.)

[34] W. Rudin. *Principles of Mathematical Analysis*. third edition, 1976. International Series in Pure and Applied Mathematics. McGraw-Hill. (Cited on pp. 3 and 10.)

[35] W. Rudin. *Real and Complex Analysis*. McGraw-Hill, third edition, 1987. (Cited on pp. 150 and 166.)

[36] R. Rudnicki and M. Tyran-Kamińska. *Piecewise Deterministic Processes in Biological Models*. Springer Briefs in Applied Sciences and Technology (Springer Briefs in Mathematical Methods). Springer, 2017. (Cited on p. 233.)

[37] R. A. Ryan. *Introduction to Tensor Products of Banach Spaces*. Springer, 2002. (Cited on p. 86.)

[38] A. D. Sokal. A really simple elementary proof of the uniform boundedness theorem. *Amer. Math. Monthly*, 118(5):450–452, 2011. (Cited on p. 167.)

[39] H. Steinhaus. Anwendungen der Funktionalanalysis auf einige Fragen der reellen Funktionentheorie. *Fundamenta Mathematica*, 1:51–81, 1929. (Cited on p. 187.)

[40] W. Tatarkiewicz. *Historia filozofii*. Państwowe Wydawnictwo Naukowe, 1958. (Cited on pp. 5 and 6.)

[41] M. E. Taylor. *Partial Differential Equations I. Basic Theory*, volume 115 of Applied Mathematical Sciences. Springer, second edition, 2011. (Cited on p. 115.)

[42] H. F. Weinberger. *A First Course in Partial Differential Equations with Complex Variables and Transform Methods*. Blaisdell 1965. (Cited on p. 115.)

[43] J. Zemánek. A simple proof of the Weierstrass–Stone theorem. *Comment. Math. Prace Mat.*, 20(2):495–497 (loose errata), 1977/78. (Cited on p. 72.)

Index

Printed in the United States
by Baker & Taylor Publisher Services